Field Theory

MONOGRAPHS AND TEXTBOOKS IN PURE AND APPLIED MATHEMATICS

1. K. Yano. Integral Formulas in Riemannian Geometry (1970)
2. S. Kobayashi. Hyperbolic Manifolds and Holomorphic Mappings (1970)
3. V. S. Vladimirov. Equations of Mathematical Physics (A. Jeffrey, editor; A. Littlewood, translator) (1970)
4. B. N. Pshenichnyi. Necessary Conditions for an Extremum (L. Neustadt, translation editor; K. Makowski, translator) (1971)
5. L. Narici, E. Beckenstein, and G. Bachman. Functional Analysis and Valuation Theory (1971)
6. D. S. Passman. Infinite Group Rings (1971)
7. L. Dornhoff. Group Representation Theory (in two parts). Part A: Ordinary Representation Theory. Part B: Modular Representation Theory (1971, 1972)
8. W. Boothby and G. L. Weiss (eds.). Symmetric Spaces: Short Courses Presented at Washington University (1972)
9. Y. Matsushima. Differentiable Manifolds (E. T. Kobayashi, translator) (1972)
10. L. E. Ward, Jr. Topology: An Outline for a First Course (1972)
11. A. Babakhanian. Cohomological Methods in Group Theory (1972)
12. R. Gilmer. Multiplicative Ideal Theory (1972)
13. J. Yeh. Stochastic Processes and the Wiener Integral (1973)
14. J. Barros-Neto. Introduction to the Theory of Distributions (1973)
15. R. Larsen. Functional Analysis: An Introduction (1973)
16. K. Yano and S. Ishihara. Tangent and Cotangent Bundles: Differential Geometry (1973)
17. C. Procesi. Rings with Polynomial Identities (1973)
18. R. Hermann. Geometry, Physics, and Systems (1973)
19. N. R. Wallach. Harmonic Analysis on Homogeneous Spaces (1973)
20. J. Dieudonné. Introduction to the Theory of Formal Groups (1973)
21. I. Vaisman. Cohomology and Differential Forms (1973)
22. B.-Y. Chen. Geometry of Submanifolds (1973)
23. M. Marcus. Finite Dimensional Multilinear Algebra (in two parts) (1973, 1975)
24. R. Larsen. Banach Algebras: An Introduction (1973)
25. R. O. Kujala and A. L. Vitter (eds). Value Distribution Theory: Part A; Part B. Deficit and Bezout Estimates by Wilhelm Stoll (1973)
26. K. B. Stolarsky. Algebraic Numbers and Diophantine Approximation (1974)
27. A. R. Magid. The Separable Galois Theory of Commutative Rings (1974)
28. B. R. McDonald. Finite Rings with Identity (1974)
29. I. Satake. Linear Algebra (S. Koh, T. Akiba, and S. Ihara, translators) (1975)
30. J. S. Golan. Localization of Noncommutative Rings (1975)
31. G. Klambauer. Mathematical Analysis (1975)
32. M. K. Agoston. Algebraic Topology: A First Course (1976)
33. K. R. Goodearl. Ring Theory: Nonsingular Rings and Modules (1976)
34. L. E. Mansfield. Linear Algebra with Geometric Applications (1976)
35. N.J. Pullman. Matrix Theory and its Applications: Selected Topics (1976)
36. B. R. McDonald. Geometric Algebra Over Local Rings (1976)
37. C. W. Groetsch. Generalized Inverses of Linear Operators: Representation and Approximation (1977)
38. J. E. Kuczkowski and J. L. Gersting. Abstract Algebra: A First Look (1977)
39. C. O. Christenson and W. L. Voxman. Aspects of Topology (1977)
40. M. Nagata. Field Theory (1977)

Field Theory

Masayoshi Nagata

Department of Mathematics
Kyoto University
Kyoto, Japan

MARCEL DEKKER, INC. New York and Basel

Library of Congress Cataloging in Publication Data

Nagata, Masayoshi, 1927-
Field theory.

(Pure and applied mathematics ; 40)
Includes index.
1. Fields, Algebraic. 2. Field extensions (Mathematics) I. Title.
QA247.N25 512'.32 76-11106
ISBN 0-8247-6466-8

MARCEL DEKKER, INC.

270 Madison Avenue, New York, New York 10016

Current printing (last digit):
10 9 8 7 6 5 4 3 2 1

PRINTED IN THE UNITED STATES OF AMERICA

PREFACE

The theory of fields has traditionally been considered a basic part of abstract algebra. In modern mathematics, however, the abundance of algebraic ideas which have been introduced has established the importance of the theory of fields not only in algebra but throughout all areas of mathematics as well. There are many topics that require discussion in the mathematics courses frequently offered in colleges and universities. Consequently, it is not unusual to have too few lecture hours devoted to the theory of fields.

In view of this, the author wished to publish a book on field theory, rich in topical variety, necessitating few prerequisites, and conveniently sized, that would allow any student to advance his study.

In this book, it is assumed that the reader is familiar with the basic definitions and results on set theory and determinants, and therefore these results are stated explicitly and without proof. In addition, some basic results on group theory and ring theory will be needed; proofs for these results are given.

Thus the main text of this volume initially consists of preliminaries on groups and rings (Chapters 1 and 2) and in subsequent chapters (Chapters 3 to 7) focuses on several important topics on fields, such as algebraic and transcendental extensions, valuations, ordered fields, and Galois theory of algebraic extensions.

M. Nagata

CONTENTS

CHAPTER 0

CONVENTIONS AND BASIC RESULTS ON SET THEORY

0.0 NOTATION

We assume that the reader is familiar with fundamental concepts in set theory such as sets, elements, subsets, the empty set, unions of sets and intersections of sets. We use the following symbols.

$\in$ $a \in M$ means that a is an element of M.

$\notin$ $a \notin M$ means that a is not an element of M.

$\subseteq$ $N \subseteq M$ means that N is a subset of M.

$\subset$ $N \subset M$ means that N is a proper subset of M, namely that $N \subseteq M$ and $N \neq M$. (Note that often in the literature $N \subset M$ means that N is a subset of M. The usage is different from ours.)

$\nsubseteq$ $N \nsubseteq M$ means that N is not a subset of M.

$\cup, \cap$ Union and intersection, respectively. Namely, the union and the intersection of sets $M_1, \ldots, M_n$ are denoted by $M_1 \cup M_2 \cup \cdots \cup M_n$ (or $\cup_{i=1}^{n} M_i$) and $M_1 \cap M_2 \cap \cdots \cap M_n$ (or $\cap_{i=1}^{n} M_i$) respectively. If a family of sets M_λ is indexed by a set Λ, then the union and the intersection of these M_λ are denoted by $\cup_{\lambda \in \Lambda} M_\lambda$ (or, simply, $\cup_\lambda M_\lambda$ or $\cup M_\lambda$) and $\cap_{\lambda \in \Lambda} M$ (or, simply, $\cap_\lambda M_\lambda$ or $\cap M_\lambda$), respectively.

$\{\ |\ \}$ When P is a condition and M is a set, the set of elements of M satisfying the condition P is denoted by $\{a \in M \mid P\}$ or simply by $\{a \mid P\}$.

$-$ When B is a subset of a set A, we denote by $A - B$ the complement of B in A (i.e., $\{a \in A \mid a \notin B\}$).

×,Π The product of sets $M_1, \ldots, M_n$, i.e., $\{(a_1, \ldots, a_n) \mid a_i \in M_i\}$, is denoted by $M_1 \times M_2 \times \cdots \times M_n$. For a family of sets M_λ ($\lambda \in \Lambda$), the product of these sets M_λ is denoted by $\Pi_{\lambda \in \Lambda} M_\lambda$.

N The set of natural numbers, i.e., $N = \{1,2,3,\ldots\}$.

0.1 MAPPINGS

A mapping f of a set M into a set N is a correspondence which associates with each element of M a single element of N. If n ($\in$N) is associated with m ($\in$M) (the circumstance is often expressed by $m \leadsto n$) by f, then n is called the image of m under f. In expressing images, two types of notation are commonly used. One is fm and the other is m^f. If M_1 is a subset of M, then $fM_1 = \{fm \mid m \in M_1\}$ or $M_1^f = \{m^f \mid m \in M_1\}$ is called the image of M_1 under f. If fM = N or $M^f = N$, then we say that f is a mapping of M onto N and that f is surjective. For a subset N' of N, $\{x \in M \mid fx \in N'$ (or $x^f \in N')\}$ is called the inverse image of N' under f and is denoted by $f^{-1}(N')$. If, for $n \in N$, $f^{-1}(\{n\})$ consists of a single element m, then m is also called the inverse image of n under f and m is denoted by $f^{-1}m$ or $m^{f^{-1}}$. If f is surjective and if every element n of N has the inverse image, then the mapping f^{-1} is well defined, which is called the inverse of f. Note that this condition is that f gives a one-one correspondence.

The projection of the product set $M_1 \times M_2 \times \cdots \times M_n$ into M_i is defined by the correspondence $(a_1,\ldots,a_n) \leadsto a_i$. Projections in the case of the product of an infinite number of sets M_λ are defined similarly. Note that projections are surjective unless some M_λ is empty.

Assume that f is a mapping of a set M into a set N and that g is a mapping of the set N into a set P. Then the composition of these mappings is defined by associating with each element m of M the image under g of the image of m under f. In case we are using notation of type m^f, the product (i.e., the result of the composition) of the mappings f and g is denoted by fg. In case we are using notation of type fm, the product is denoted by gf. The reason is that, under the notation, we have $(m^f)^g = m^{fg}$ and $g(fm) = (gf)m$.

For each set M, there is defined the cardinality of M, which is denoted by #(M). If M consists only of a finite number of elements, then M is called a finite set, and #(M) is the number of elements of M. [If M is the empty set, then #(M) = 0.] The cardinality of the set $\mathbb{N}$ of natural numbers is said to be countably infinite. (Countable means finite or countably infinite.) The cardinality of the set of real numbers is called the cardinality of continuum. Multiplication of cardinalities is defined by $\#(M) \times \#(N) = \#(M \times N)$. In the case of finite cardinalities, the multiplication coincides with that of numbers. But, as for infinite cardinalities, the multiplication is quite different from the case of numbers; cf. Theorem 0.1.3 below. In general, we define that #(M) = #(N) if and only if there is a one-one correspondence between M and N. We define also that $\#(M) \geq \#(N)$ if and only if there is a subset M' of M such that #(M') = #(N). Hence $\#(M) > \#(N)$ means that $\#(M) \geq \#(N)$ and $\#(M) \neq \#(N)$. Under the definition, we see obviously that $\#(M) \geq \#(N)$ and $\#(N) \geq \#(L)$ imply $\#(M) \geq \#(L)$. Furthermore, the following theorem holds.

THEOREM 0.1.1 (Bernstein) If M and N are sets, then either $\#(M) \geq \#(N)$ or $\#(N) \geq \#(M)$. If both of these inequalities hold, then #(M) = #(N).

Some other important theorems on cardinality are:

THEOREM 0.1.2 For an arbitrary set M, let S(M) be the set of all subsets of M. Then it holds that

$$\#(S(M)) > \#(M)$$

If #(M) is countably infinite, then #(S(M)) is the cardinality of continuum. In particular, the set of real numbers is not countable.

THEOREM 0.1.3 If M is an infinite set, then

$$\#(M) = \#(M \times M)$$

We omit the proofs of these results.

0.2 ORDERED SET

If a relation $\geq$ is defined on a set M (i.e., for each pair of elements a, b of M, it is well determined whether $a \geq b$ or not) and if the relation satisfies the following three conditions, then we say that $\geq$ is an order and that M is an ordered set:

(1) $a \geq a$ for every $a \in M$ (reflexive property).

(2) $a \geq b$, $b \geq a$ imply $a = b$ (asymmetric property).

(3) $a \geq b$, $b \geq c$ imply $a \geq c$ (transitive property).

$a > b$ means $a \geq b$ and $a \neq b$, and we say usually in such a case that a is larger (or greater) than b, or that b is smaller than a. $a < b$ (or $a \leq b$) means $b > a$ (or $b \geq a$, respectively). The notion of an ordered set is actually a pair of a set M and an order $\geq$ defined on it. Therefore, if a set M has two orders $\geq$ and $\propto$, then M with $\geq$ and M with $\propto$ are distinct from each other. Therefore in order to express that M is an ordered set with order $\geq$, we often say that $(M,\geq)$ is an ordered set.

For a set N, let $S(N)$ be the set of subsets of N. Then by the containment relation $\supseteq$, $(S(N)),\supseteq)$ is an ordered set. When we deal with a subset of $S(N)$, we understand it as a subset of this ordered set.

Assume that $(M,\geq)$ is an ordered set. Then we can define another order $\propto$ on the same set M by defining

$$a \propto b \qquad \text{if and only if } b \geq a$$

This new order is called the dual of the former.

We say that an ordered set $(M,\geq)$ is linearly ordered if for each pair of elements a, b of M, it holds either $a \geq b$ or $b \geq a$.

A subset of an ordered set is obviously an ordered set, and a subset of a linearly ordered set is a linearly ordered set.

Let S be a subset of an ordered set $(M,\geq)$, and let a be an element of S. (i) If there is no b in S which is greater than a, then we say that a is a maximal element of S. (ii) If a is larger than any other elements of S, then we say that a is the largest (or greatest) element of S. An element y of M is called a lower bound of S if y is smaller than any element (except y in case y is in S) of S. If the set of lower bounds of S has the largest element, say z, then z is called the infimum of S and is denoted by inf S.

Considering the dual (hence, interchanging large and small), we define minimal elements, the smallest element, upper bound and the supremum of S. This last is denoted by sup S.

Note that S may not have any of minimal element, maximal element, infimum, supremum, etc. Even if S has the infimum, inf S may not be an element of S.

We say that the maximum condition (or the minimum condition) is satisfied by an ordered set M if every nonempty subset S of M has at least one maximal element (or minimal element, respectively). We say that the ascending chain condition holds in an ordered set $(M,\geq)$ if, for every ascending chain $a_1 \leq a_2 \leq \cdots \leq a_n \leq \cdots$ of infinite length in M, there is a natural number N such that $a_m = a_N$ for every $m > N$. This condition is equivalent to the statement that there is no properly ascending chain of infinite length in M. We define the descending chain condition in the dual way. Then we have:

THEOREM 0.2.1 The maximum conditon is equivalent to the ascending chain condition. Similarly, the descending chain condition is equivalent to the minimum condition.

PROOF: If there is a properly ascending chain $a_1 < a_2 < \cdots < a_n < \cdots$ of infinite length, then the set of all a_i has no maximal element. Conversely, if S is a nonempty subset which has no maximal element, then starting with an arbitrary element a_1 of S we have a properly ascending chain $a_1 < a_2 < \cdots < a_n$ of element of S. Then, since S has no maximal element, there is a_{n+1} $(\in S)$ which is larger than a_n. Thus we have such a chain of infinite length. This proves the theorem, taking account of the dual. QED

If the minimum condition holds in a linearly ordered set M, then we say that M is a well-ordered set and that the order of M is a well-order.

Let $(M_1,\geq)$, .., $(M_n,\geq)$ be ordered sets. Then we can define a new order $\geq$ in the product set $M_1 \times M_2 \times \cdots \times M_n$ as follows:

$(a_1,\ldots,a_n) > (b_1,\ldots,b_n)$ if and only if there is one i such that $a_i > b_i$ and such that, for every j smaller than i, it holds that $a_j = b_j$.

This new order is called the lexicographical order.

In closing this section, we recall a well-known and important theorem, the Zorn lemma. For this purpose, we define the notion of an inductive set to be a nonempty ordered set M in which every nonempty well-ordered subset S has the supremum sup S in M. Now:

ZORN LEMMA If M is an inductive set, then there is a maximal element of M.

0.3 CLASSIFICATION

A classification on a set M means to express M as a disjoint union of nonempty subsets C_λ (λ runs through a set Λ), namely, (i) $M = \cup_\lambda C_\lambda$, (ii) if $C_\lambda \cap C_\mu$ is not empty then $C_\lambda = C_\mu$, and (iii) each C_λ is not empty. Each C_λ is called a class, and if a belongs to C_λ then a is called a representative of C_λ.

If a realtion $\equiv$ defined on a set M satisfies the following three conditions, then we say that $\equiv$ is an equivalence relation:

(1) $a \equiv a$ for every $a \in M$ (reflexive property).

(2) $a \equiv b$ implies $b \equiv a$ (symmetric property).

(3) $a \equiv b$, $b \equiv c$ imply $a \equiv c$ (transitive property).

As is well known, the notion of classification is closely related to that of equivalence relation. Namely,

THEOREM 0.3.1 Assume that a set M is the union of nonempty subsets C_λ ($\lambda \in \Lambda$). Define a relation $\equiv$ by that $a \equiv b$ if and only if there is λ such that both a and b are in C_λ. Then the relation $\equiv$ is an equivalence relation if and only if these C_λ give a classification on M.

The proof is easy.

EXERCISE 0.2

1. (<u>(Mathematical) induction</u>) Let M be an ordered set satisfying the minimum condition. Assume that there is given a statement P_a for each element a of M. Then all P_a $(a \in M)$ are true if the following is true :

   If $a \in M$ and if P_b is true for every b such that $a > b \in M$, then P_a is true.
2. Let W be a well-ordered set and let $\{M_w \mid w \in W\}$ be a set of ordered sets indexed by W. Generalize the definition of the lexicographical order to the product set ΠM_w.
3. Let $M_1, \ldots, M_n$ be ordered sets and consider the lexicographical order on $M_1 \times \cdots \times M_n$.
   (i) Prove that if $M_1, \ldots, M_n$ are linearly ordered, then $M_1 \times \cdots \times M_n$ is also linearly ordered.
   (ii) Prove that if $M_1, \ldots, M_n$ are well-ordered, then $M_1 \times \cdots \times M_n$ is also well-ordered.

CHAPTER 1

GROUPS

In this chapter, we recall some basic notions on groups which we need later. We assume that the readers are familiar with some elementary properties of rational integers.†

1.1 GROUPS AND SEMIGROUPS

A binary operation, often simply called an operation, on a set M is a mapping ψ of the product set $M \times M$ into M. The operation is said to be commutative if $\psi(a,b) = \psi(b,a)$ for every $(a,b) \in M \times M$. A binary operation is often called either a multiplication or an addition. If ψ is called a multiplication, then $\psi(a,b)$ is called the product of a and b; it is quite common that $\psi(a,b)$ is denoted by ab in this case, and we follow the custom. If ψ is called an addition, then $\psi(a,b)$ is called the sum of a and b and is denoted by $a + b$; it is quite common to assume that an addition is commutative, and we follow this too.

A semigroup is a set G with a binary operation, say a multiplication, satisfying associativity, i.e.,

$$a(bc) = (ab)c \qquad \text{for arbitrary } a,b,c \in G$$

If a semigroup G has an element e such that $ea = ae = a$ for every $a \in G$, then such e is unique (see Proposition 1.1.1

† As we shall study later in Chap. 3, there is the notion of algebraic integers. Therefore, integers $0, \pm 1, \pm 2, \ldots$ are called rational integers.

below) and is called the *identity* of G; it is denoted by 1 or more explicitly by 1_G. Assume that a semigroup G has the identity. If, for an element a of G, there is an element a' such that aa' = a'a = 1, then such a' is unique (Proposition 1.1.1) and is called the *inverse* of a; it is denoted by a^{-1}. An element a of G having its inverse is called an *invertible* element of G.

A semigroup is called a *group* if (1) it is a semigroup with identity and (2) every element is invertible in the semigroup.

If the operation of a semigroup (or a group) is commutative, then the semigroup (or the group) is said to be *commutative*. A commutative group is sometimes called an *abelian* group.

If a set M is a group with addition +, then the identity is called *zero*, denoted by 0, and the inverse of an element a is called the *minus* of a and is denoted by -a. M itself is called a *module* or an *additive group*.

PROPOSITION 1.1.1 In a semigroup G with identity, there is only one identity; for each $a \in G$, the inverse a^{-1} is unique if it exists, and then $(a^{-1})^{-1} = a$.

PROOF: If 1' is another identity, the 1 = 11' = 1'. If a' is another inverse of a, then $a^{-1} = a^{-1}(aa') = (a^{-1}a)a' = a'$. a is an inverse of a^{-1}, and hence $(a^{-1})^{-1} = a$. QED

If a subset H of a group G forms a group under the restriction of the operation of G to H, then we say that H is a *subgroup* of G. *Submodules* and *subsemigroups* are defined similarly.

If H_λ $(\lambda \in \Lambda)$ are subgroups of a group G, then the intersection of these H_λ is a subgroup. Therefore, when a subset S of a group G is given, the intersection D of all subgroups containing S is the smallest subgroup containing S. D is therefore called the group *generated* by S and is denoted by $<S>$. If a subgroup H is generated by a subset S, then S is called a system of *generators* for H. A group generated by a single element is called a *cyclic* group. When H and K are subgroups of a group G, the subgroup generated by $H \cup K$ is denoted by $H \vee K$. Similar notation is employed for the subgroup generated by many subgroups.

THEOREM 1.1.2 For a nonempty subset H of a group G, each of the following two is necessary and sufficient for H to be a subgroup:

(1) $ab^{-1} \in H$ for arbitrary $a,b \in H$.

(2) $a^{-1}b \in H$ for arbitrary $a,b \in H$.

PROOF: Necessity is obvious. Assume that (1) holds good. Let $a \in H$. It holds that $a,a \in H$ which implies $1 = aa^{-1} \in H$ and then $1,a \in H$ which implies $a^{-1} \in H$. Now, if $a,b \in H$, then $a,b^{-1} \in H$; hence $ab = a(b^{-1})^{-1} \in H$. Sufficiency of (2) is similar. QED

If H, K are subsets of a group G, we denote by HK the set $\{hk \mid h \in H, k \in K\}$ (in case of an additive group, $H + K = \{h + k \mid h \in H, k \in K\}$). Hence, in particular, when H is a subgroup and a is an element of G, we denote by Ha the set $\{ha \mid h \in H\}$ and by aH the set $\{ah \mid h \in H\}$ (in the additive case, $H + a = \{h + a \mid h \in H\}$). Ha (or aH) is called the right (or left, respectively) residue class (or coset) of a modulo H. The set of all right (or left) residue classes of elements of G modulo H is denoted by $H\backslash G$ (or G/H, respectively).

PROPOSITION 1.1.3 Under the circumstances above,

(1) $aH = bH$ if and only if $b^{-1}a \in H$.

(2) $Ha = Hb$ if and only if $ab^{-1} \in H$.

(3) $aH \neq bH$ if and only if $aH \cap bH$ is empty.

(4) $Ha \neq Hb$ if and only if $Ha \cap Hb$ is empty.

PROOF: If $aH = bH$, then $a = bh$ with $h \in H$; hence $b^{-1}a \in H$. Conversely, if $b^{-1}a \in H$, then $a \in bH$ and $aH \subseteq bH$. Similarly, $bH \subseteq aH$ because $b = ah^{-1}$. (2) is similar. If $c \in aH \cap bH$, then by (1) we have $aH = cH = bH$. Since the converse is obvious, we have (3). (4) is similar. QED

The result stated above shows that G/H and $H\backslash G$ are giving classifications on G.

For a group G, $\#(G)$ is called the order of G. If $\#(G)$ is finite, we call G a finite group. For an element a of G, the order of the cyclic group $\langle a \rangle$ is called the order of the element a.

(These notions of orders are completely different from the notion of an order in the case of an ordered set.) When H is a subgroup of a group G, $\#(G/H)$, which is equal to $\#(H\backslash G)$ as will be shown below, is called the <u>index</u> of H in G.

THEOREM 1.1.4 Let H be a subgroup of a group G. Then $\#(G/H) = \#(H\backslash G)$. If $\#(G)$ is finite, then $\#(G) = \#(H) \times \#(G/H)$.

PROOF: The mapping $a \rightsquigarrow a^{-1}$ induces a one-one correspondence between G/H and $H\backslash G$ $(Ha \rightsquigarrow a^{-1}H)$ as is easily seen. If $\#(G)$ is finite, then $\#(H)$ is finite and each Ha or aH consists exactly of $\#(H)$ elements. QED

COROLLARY 1.1.5 In a finite group G, the order of a subgroup H, the index of H and the order of an element are all divisors of $\#(G)$.

We add here one remark on cyclic groups.

LEMMA 1.1.6 Assume that an element a of a group G has order n (finite). If a natural number m is relatively prime to n, then $\langle a\rangle = \langle a^m\rangle$.

PROOF: It is obvious that $\langle a^m\rangle \subseteq \langle a\rangle$. There are integers c, d such that $cm + dn = 1$.† Then $(a^m)^c = a^{cm+dn} = a$. QED

One of the important examples of groups is a permutation group. For a given set M, let $S(M)$ be the set of mappings σ of M onto itself such that each σ gives a one-one correspondence of M with M itself. Then $S(M)$ forms a group under the composition of mappings. This group is called the <u>symmetric group</u> on M. A subgroup of this is called a <u>permutation group</u> on M. If $\#(M) = \#(N)$, then the symmetric groups $S(M)$ and $S(N)$ are isomorphic to each other. Therefore, if $\#(M) = n$ (finite), then $S(M)$ is called the <u>symmetric group of degree</u> n, and a subgroup is a <u>permutation group of degree</u> n. Note that in this case $\#(S(M)) = n!$.

† Though this is elementary knowledge of numbers, it would be noteworthy that this follows actually from the fact that the ring of rational integers is a Euclid ring; see Sec. 2.6.

1.2 NORMAL SUBGROUPS AND HOMOMORPHISMS

A mapping ψ of a group G into a group G' is called a homomorphism if it holds that $\psi(ab) = (\psi a)(\psi b)$ for arbitrary $a,b \in G$. In this case, $\{x \in G \mid \psi x = 1\}$ is called the kernel of ψ. If the kernel consists only of the identity, then the homomorphism is called an injection or said to be injective. On the other hand, if ψ is a homomorphism of G onto G', we say that ψ is a surjection or that ψ is surjective. (These words, injection, surjection, etc., have wider use.) The set of all homomorphisms of a group G into a group G' is denoted by $\mathrm{Hom}(G,G')$. If a homomorphism ψ is injective and at the same time surjective, then we say that ψ is an isomorphism. If there is an isomorphism of a group G onto a group G', then we say that G is isomorphic to G', and we write $G \cong G'$.

An isomorphism of a group G to itself is called an automorphism of G. The set of all automorphisms of G is called the automorphism group of G and is denoted by $\mathrm{Aut}\, G$, whose group structure is defined as a subgroup of the symmetric group on G. For elements a, x of the group G, we often denote by x^a the element $a^{-1}xa$. Then $x \rightsquigarrow x^a$ (for a fixed a) defines an automorphism of G. This is called the inner automorphism defined by a. The set of all inner automorphisms of G forms a normal subgroup of $\mathrm{Aut}\, G$ and is called the inner automorphism group of G. Two subsets H and K of G are said to be conjugate to each other if there is an inner automorphism which sends H to K, i.e., if there is an element a of G such that $a^{-1}Ha = K$. A subgroup N of G is called a normal subgroup if N itself is the only conjugate of N. In this case, $aN = Na$ for every $a \in G$; hence we call Na the residue class of a module N and often express Na by $a \bmod N$.

THEOREM 1.2.1 If N is a normal subgroup of a group G, then G/N form a group under the multiplication defined by $(aN)(bN) = abN$. The mapping $a \rightsquigarrow aN$ defines a surjection ψ of G to G/N.

This group G/N is called the residue class group of G modulo N and the surjection ψ is called the natural homomorphism of G to G/N.

PROOF: As subsets of G, we have $(aN)(bN) = abNN = abN$, and the assertions follow easily. QED

THEOREM 1.2.2 If ψ is a homomorphism of a group G into a group G', then the kernel N of ψ is a normal subgroup of G and the correspondence $aN \rightsquigarrow \psi a$ is a one-one correspondence between G/N and ψG; hence it gives an injection of G/N into G'. Furthermore, for each subgroup H' of ψG, $\psi^{-1}(H')$ is a subgroup of G containing N. This correspondence $H' \rightsquigarrow \psi^{-1}(H')$ gives a one-one correspondence between subgroups of ψG and subgroups of G containing N. Under the correspondence, normal subgroups correspond to normal subgroups.

PROOF: If $a,b \in \psi^{-1}(H')$, then $\psi(a^{-1}b) = (\psi a)^{-1}(\psi b) \in H'$; hence $\psi^{-1}(H')$ is a subgroup of G by Theorem 1.1.2. It is obvious that $\psi^{-1}(H')$ contains N. If H' is a normal subgroup of G, then for arbitrary $x \in G$ and $a \in \psi^{-1}(H')$, $\psi(x^{-1}ax) = (\psi x)^{-1}(\psi a)(\psi x) \in H'$, and we see that $\psi^{-1}(H')$ is normal. Considering the case where $H' = \{1\}$, we see that N is normal. Now, $x \in \psi^{-1}(\psi y)$ $(y \in G)$ if and only if $\psi x = \psi y$, i.e., $\psi(x^{-1}y) = 1$, namely, $xN = yN$. Therefore we see that $aN \rightsquigarrow \psi a$ gives a one-one correspondence which is an isomorphism of G/N to ψG. If H is a subgroup of G containing N, then that $b \in \psi^{-1}(\psi H)$ implies that $bN = hN$ with $h \in H$, hence $\psi^{-1}(\psi H) \subseteq HN = H$ and thus $\psi^{-1}(\psi H) = H$. Obviously ψH is a subgroup of ψG, and we see that $H' \rightsquigarrow \psi^{-1}(H')$ gives a one-one correspondence as stated above. If H is a normal subgroup of G, then we see easily that ψH is a normal subgroup of ψG, and we complete the proof. QED

Note that the two theorems stated above show that a subset of a group is a normal subgroup if and only if it is the kernel of some homomorphism of the group.

Let ψ be a homomorphism of a group G into a group G', let N be the kernel of ψ, and let H be a subgroup of G. Then the restriction $\psi|_H$ of ψ to H is a homomorphism of H into G' and the kernel of $\psi|_H$ is obviously $H \cap N$. Therefore we have the following important theorem which is called the <u>isomorphism theorem</u>.

THEOREM 1.2.3 Under the assumption stated above, $H/(H \cap N)$ is regarded as a subgroup of G/N, identifying hN $(h \in H)$ of G/N with $h(H \cap N)$ of $H/(H \cap N)$. In particular, if furthermore $G = HN$, then $H/(H \cap N)$ is identified with G/N.

1.3 SOLVABILITY

For elements a, b of a group G, the element $a^{-1}b^{-1}ab$ is called the <u>commutator</u> of a and b and is denoted by $[a,b]$. The name is after the fact that $ab = ba[a,b]$. When H and K are subgroups of G, $[H,K]$ denotes the subgroup generated by all of $[h,k]$ $(h \in H, k \in K)$. $[H,K]$ is called the <u>commutator subgroup</u> of H and K and $[G,G]$ is called the <u>commutator group</u> of G or the <u>derived group</u> of G.

THEOREM 1.3.1 If H and K are normal subgroups of a group G, then $[H,K]$ is a normal subgroup of G and is contained in $H \cap K$.

PROOF: Since $x^{-1}[a,b]x = [x^{-1}ax, x^{-1}bx]$, we see that $x^{-1}\{[h,k] \mid h \in H, k \in K\}x = \{[h,k] \mid h \in H, k \in K\}$ for every $x \in G$. Therefore $[H,K]$ is a normal subgroup. $[h,k] = h^{-1}k^{-1}hk$, and $h^{-1}k^{-1}h \in K$ for $k \in K$. Hence $[H,K] \subseteq K$. Similarly, $[H,K] \subseteq H$. QED

COROLLARY 1.3.2 For a group G, we define D_i and Z_i by $G = D_0 = Z_0$, $D_i = [D_{i-1}, D_{i-1}]$, $Z_i = [Z_{i-1}, G]$ for $i = 1, 2, \ldots$. Then these D_i and Z_j are normal subgroups of G.

The sequence $D_0 \supseteq D_1 \supseteq D_2 \supseteq \cdots$ is called the <u>derived series</u> of G and the sequence $Z_0 \supseteq Z_1 \supseteq Z_2 \supseteq \cdots$ is called the <u>descending central series</u> of G. If there is one i such that $D_i = \{1\}$, then G is called a <u>solvable</u> group. If there is one i such that $Z_i = \{1\}$, then G is called a <u>nilpotent</u> group. A nilpotent group is solvable.

If S is a subset of a group G, and if H is a subgroup of G, we denote by $Z_H(S)$ the set $\{h \in H \mid hs = sh$ for every $s \in S\}$ which is called the <u>centralizer</u> of S in H. $Z_G(G)$ is called the <u>center</u> of G and is denoted by $Z(G)$. A sequence $C_0 = \{1\} \subseteq C_1 \subseteq$

$C_2 \subseteq \cdots$ of subgroups of G is called the <u>ascending central series</u> of G if each C_i/C_{i-1} is the center of G/C_{i-1}.

THEOREM 1.3.3 A group G is nilpotent if and only if G appears in its ascending central series.

PROOF: C_i and Z_j being as above, we first assume that $Z_n = \{1\}$ for an n, and we shall prove that $C_n = G$ by induction on n. If $n = 1$, then G is commutative and the assertion is obvious. We assume that $n > 1$. $[G, Z_{n-1}] = Z_n = \{1\}$ means that $Z_{n-1} \subseteq C_1$. We consider the natural homomorphism ψ of G onto G/C_1. Then the descending central series of ψG is $\psi Z_0 \supseteq \psi Z_1 \supseteq \cdots$ (cf. Exercise 1.3.2). Since $\psi Z_{n-1} = \{1\}$, and since $\psi C_1 \subseteq \psi C_2 \subseteq \cdots$ is the ascending central series of ψG, we see that $\psi C_n = \psi G$, i.e., $C_n = G$, by our induction assumption. Conversely, assume that $C_n = G$, and we shall prove that $Z_n = \{1\}$ by induction on n. If $n = 1$, then G is commutative and $Z_1 = \{1\}$. Assume that $n > 1$ and consider as above. Then our induction assumption shows that $\psi Z_{n-1} = \{1\}$, and hence that $Z_{n-1} \subseteq C_1$. Therefore $Z_n = \{1\}$. QED

A finite group is called a <u>p-group</u> if p is a prime number and if the order of the group is a power of p.

THEOREM 1.3.4 If G is a p-group, then G is nilpotent.

PROOF: For each $a \in G$, let B_a be the set of b $(\in G)$ which are conjugate to a. For $x, y \in G$, $x^{-1}ax = y^{-1}ay$ if and only if $xy^{-1} \in Z_G(a)$. Therefore $\#(B_a) = \#(G/Z_G(a))$, which is a power of p by Corollary 1.1.5. Hence $\#(B_a)$ is not divisible by p if and only if $\#(B_a) = 1$, i.e., $a \in Z(G)$. On the other hand, these B_a give a classification on G, and hence the sum of $\#(B_a)$ is the order of G. Therefore $\#(Z(G)) \geq p$ unless $\#(G) = 1$. Thus, in the ascending central series $C_0 \subseteq C_1 \subseteq \cdots$, $C_n = C_{n+1}$ implies that $C_n = G$. Thus we see the nilpotency of G. QED

1.4 SYLOW THEOREM

We say that a group G <u>acts</u> on a set E if for each $(x,a) \in G \times E$ the product xa is defined so that (i) $xa \in E$ and (ii) $(xy)a = x(ya)$ for arbitrary $x, y \in G$, $a \in E$. $Ga = \{xa \mid x \in G\}$ is called the orbit (or G-orbit) of a. $\{x \in G \mid xa = a\}$ is a subgroup of G and is called the stabilizer for a. If $xa = a$ for all x in G, then a is called an invariant (or G-invariant). The set of G-invariant in E is usually denoted by E^G.

LEMMA 1.4.1 Under the notation as above, (1) G-orbits in E give a classification on E, (2) $xa = ya$ $(x,y \in G,\ a \in E)$ if and only if $x^{-1}y$ is in the stabilizer S_a for a and therefore $\#(Ga) = \#(G/S_a)$.

The proof is easy and we omit it.

LEMMA 1.4.2 If a p-group P acts on a finite set E, then $\#(E) \equiv \#(E^P) \pmod{p}$.†

PROOF: For each $a \in E$, let S_a be the stabilizer for a. Then $\#(Pa) = \#(P/S_a)$ by the lemma above. Therefore $\#(Pa) \equiv 0 \pmod{p}$ if and only if $a \notin E^P$. QED

Now we come to the famous Sylow theorem:

THEOREM 1.4.3 Let p be a prime number and let G be a finite group. Assume that $\#(G) = p^e q$ (q and e being rational integers with q prime to p). Then (1) there is a subgroup S of G such that $\#(S) = p^e$; (2) if S' is another such S, then S' is conjugate to S; (3) $\#(\{x^{-1}Sx \mid x \in G\}) \equiv 1 \pmod{p}$; and (4) if a subgroup H of G is a p-group, then H is contained in some conjugate of S.

Such an S is called a p-Sylow subgroup of G.

PROOF: Set $E = \{X \subseteq G \mid \#(X) = p^e\}$. Then the coincidence of the numbers of combinations with binomial coefficients implies that $\#(E)$ coincides with the coefficient of X^{p^e} in the expansion of $(X + 1)^{p^e q}$. But, $(X + 1)^{p^e q} \equiv (X^{p^e} + 1)^q \pmod{p}$.† Since q is

† Two (rational) integers a and b are said to be congruent to each other modulo an integer n if $a - b$ is divisible by n. We express the situation by $a \equiv b \pmod{n}$. Furthermore, this definition is extended to the case of polynomials with integral coefficients in such a way that $f(x) \equiv g(x) \pmod{n}$ if all the coefficients of $f(x) - g(x)$ are divisible by n.

the coefficient of X^{p^e} in the expansion of $(X^{p^e}+1)^q$, we see that $\#(E) \equiv q \pmod{p}$. In particular, $\#(E)$ is not divisible by p.

Now, for arbitrary $y \in G$, $X \in E$, we see that $yX \in E$, and therefore a natural action of G on E is defined. Since orbits give a classification on E and since $\#(E)$ is not a multiple of p, there is an $X \in E$ such that $\#(GX)$ (as a subset of E) is not a multiple of p. We fix such an X and let S be the stabilizer for X. Since $\#(GX) = \#(G/S)$ (Lemma 1.4.1) and since this is not a multiple of p, we see that $\#(S)$ must be a multiple of p^e (Theorem 1.1.4). On the other hand, if $a \in X$, then $Sa \subseteq X$ and therefore $\#(S) \leq \#(X) = p^e$. Therefore S is a Sylow subgroup. Consider H in (4). H acts naturally on G/S. Since $\#(G/S)$ is not a multiple of p, there is an H-invariant cS in G/S, i.e., $hcS = cS$ for every $h \in H$ (Lemma 1.4.2). This means that $h \in cSc^{-1}$. This shows (4). (2) follows from (4). As for (3), Let N be the normalizer of S (cf. Exercise 1.1.4). Then the number of conjugates of S coincides with $\#(G/N)$ (cf. Exercise 1.4.1). Now, consider the natural action of S on G/S. Lemma 1.4.2 asserts that $\#(G/S) \equiv \#((G/S)^S) \pmod{p}$. On the other hand, $xS \in (G/S)^S$ if and only if $xSx^{-1} = S$, and hence $\#((G/S)^S) = \#(N/S)$. Therefore we see that $\#(G/N) = \#(G/S)/\#(N/S) \equiv 1 \pmod{p}$. QED

1.5 DIRECT PRODUCT

We are going to deal with direct products of a finite number of groups (for infinite case, see Exercise 1.5.1). We begin with a constructive definition:

Let $H_1, \ldots, H_n$ be groups (with multiplication). In the product set $H_1 \times \cdots \times H_n$, we define a multiplication by $(a_1,\ldots,a_n)(b_1,\ldots,b_n) = (a_1b_1,\ldots,a_nb_n)$. Then, as one sees easily, this product set forms a group where the identity is $(1,\ldots,1)$ and

† For polynomials f, g with integral coefficients, we see easily that $(f+g)^p \equiv f^p + g^p \pmod{p}$ (provided of course that p is a prime number) in view of the nature of binomial coefficients. Therefore, by virtue of induction, we have $(f+g)^{p^e} \equiv f^{p^e} + g^{p^e}$ (mod p).

$(a_1,\ldots,a_n)^{-1} = (a_1^{-1},\ldots,a_n^{-1})$. This new group is called the direct product of $H_1,\ldots,H_n$ and is denoted by $H_1 \times \cdots \times H_n$. In connection with the structural definition to be given below, we often identify each element h_i $(\in H_i)$ with $(a_1,\ldots,a_n)$ such that $a_i = h_i$ and other a_j are all identities. Then each H_i becomes a normal subgroup of the direct product, and the direct product is generated by these $H_1, \ldots, H_n$.

Next, we give a structural definition. A group G is said to be a direct product of its subgroups $N_1, \ldots, N_m$ if the following three conditions hold good:

(i) $G = N_1 \cdots N_m$; (ii) if $i \neq j$, then $[N_i,N_j] = 1$ (i.e., elements of N_i commute with elements of N_j); and (iii) for each $i = 2, \ldots, m$, $N_1 \cdots N_{i-1} \cap N_i = \{1\}$.

PROPOSITION 1.5.1 Conditions (i) to (iii) are equivalent to: (1) $N_1, \ldots, N_m$ are normal subgroups of G, and (2) each element of. G is expressed in one and only one way in the form $a_1 a_2 \cdots a_m$ with $a_i \in N_i$.

PROOF: Assume that (1) and (2) hold. If $1 \neq a \in N_1 \cdots N_{i-1} \cap N_i$, then $a = b_1 \cdots b_{i-1}$, against (2), and hence (iii) holds. Since each N_i is normal, we have $[N_i,N_j] \subseteq N_i \cap N_j$ (Theorem 1.3.1) and we have (ii) by virtue of (iii). Validity of (i) is obvious. Conversely, assume that (i) to (iii) hold. By (i) and (ii), we see that each N_i is normal. (i) gives the possibility of expressing elements of G as stated in (2); uniquess of the expression follows from (iii) quite similarly to the begining part of this proof. QED

Note that this proposition shows that our two definitions of direct product are substantially the same by virtue of the identification of H_i with a subgroup of the direct product. Direct product of modules are often called direct sums, and the symbol $\oplus$ is used instead of $\times$.

1.6 A LEMMA CYCLIC GROUPS

Later in Chapter 3, we need the following lemma:

LEMMA 1.6.1 Let G be a commutative finite group. If it holds that $\#(\{x \in G \mid x^m = 1\}) \leq m$ for every natural number m, then G is cyclic.

Though we can give a quite elementary proof of the assertion, the writer wants to give a proof of a stronger result as an application of the Sylow theorem. We begin with a lemma:

LEMMA 1.6.2 If $H = \langle a \rangle$ and $K = \langle b \rangle$ are finite cyclic groups and if $\#(H)$ is relatively prime to $\#(K)$, then the direct product $H \times K$ is cyclic.

PROOF: One sees easily that the order of the element (a,b) in $H \times K$ is $\#(H) \times \#(K)$, and hence $H \times K = \langle (a,b) \rangle$. QED

THEOREM 1.6.3 Let G be a finite group of order $g = p_1^{e_1} \cdots p_n^{e_n}$ with mutually distinct prime numbers $p_1, \ldots, p_n$ and natural numbers $e_1, \ldots, e_n$. For each natural number s, we set $E_s = \{x \in G \mid x^s = 1\}$. If $\#(E_s) \leq s$ holds for each $s = p_i^m$ $(i = 1, \ldots, n;\ m = e_i - 1, e_i)$, then G is cyclic.

PROOF: At first, we want to show that a p_i-Sylow subgroup S_i of G is normal. All the p_i-Sylow subgroups are contained in E_s for $s = p_i^{e_i}$. Since $\#(E_s) \leq s$, we see that S_i is the unique p_i-Sylow subgroup. Therefore S_i is normal. This conclusion implies that G is the direct product of $S_1, \ldots, S_n$ (in view of the structural definition of direct products). Thus, by virtue of Lemma 1.6.2, we have reduced the assertion to the case where G is a p-group and $\#(G) = p^e$. Let a be an element of G which is not in E_s with $s = p^{e-1}$. Then the order of a must be p^e and $G = \langle a \rangle$. QED

EXERCISE 1.1

1. Show that, for each natural number n, there is a cyclic group of order n.
2. If the order of a finite group G is a prime number, then G is cyclic. Prove this.
3. Give an example of noncyclic group of order 6.
4. If S is a subset of a group G, then $N(S) = \{a \in G \mid aS = Sa\}$ is called the <u>normalizer</u> of S in G. Prove that $N(S)$ is a subgroup of G.
5. If G is a cyclic group of order n and if d is a divisor of n, then there is one and only one subgroup H of order d in G. Prove this and look for the structure of H.

EXERCISE 1.2

1. If $N_1, \ldots, N_m$ are normal subgroups of a group G, then (i) the intersection of them is a normal subgroup of G, and (ii) the subgroup generated by the union of them is a normal subgroup of G and coincides with $N_1 \cdots N_m$. Prove these.
2. If N is a normal subgroup and H is a subgroup of a group G, then $N \cap H$ is a normal subgroup of H. Prove this.
3. Let G and H be finite cyclic groups of orders g and h, respectively. Let d be the greatest common measure of g and h. Let K be the set of homomorphism of G intc H. Compute $\#(K)$.
4. If H is a subgroup of a group G and if $\#(G/H) = 2$, then H is a normal subgroup of G. Prove this.

EXERCISE 1.3

1. Prove that a subgroup N of a group G contains $[G,G]$ if and only if N is a normal subgroup such that G/N is commutative.
2. Prove that if ψ is a homomorphism of a group G into a group G', and if H, K are subgroups of G, then $\psi[H,K] = [\psi H, \psi K]$.
3. Show by an example that the conditions (i) H is a normal subgroup of a group G and (ii) K is a normal subgroup of H do not imply that K is a normal subgroup of G.

4. The following three conditions are equivalent to each other:
   (i) G is a solvable group.
   (ii) The group G has a normal subgroup N such that both G/N and N are solvable.
   (iii) In the group G, there are subgroups $G_1, \ldots, G_n = \{1\}$ such that each G_i is a normal subgroup of G_{i-1} (where $G_0 = G$) and G_{i-1}/G_i is commutative.

   If the group G is a finite group, then these are equivalent to:
   (iv) There are such G_i as in (iii) with additional condition that each G_{i-1}/G_i is a cyclic group of prime order.

   Prove these.

EXERCISE 1.4

1. Let S be a subset of a group G and let N be the normalizer of S. Let E be the set of conjugates of S. Prove that $\#(E) = \#(G/N)$.
2. We say that (H,K) is a <u>Sylow pair</u> if H is a subgroup of a group K and if it enjoys the following property: If a subgroup H' of K is isomorphic to H, then H' is conjugate to H in K.

   Assume that G is a group and H and K are subgroups such that (H,K) is a Sylow pair. Let N(K) and N(H) be normalizers of K and H, respectively, in G.

   Show that $N(K) = (N(H) \cap N(K)K$.
3. Let S be a Sylow subgroup of a finite group G and let N be the normalizer of S. Prove that if a subgroup K contains N, then the normalizer of K coincides with K.

EXERCISE 1.5

1. Let $\{G_\lambda \mid \lambda \in \Lambda\}$ be a set of groups. Then the <u>direct product</u> of them is defined to be the group defined by the product set ΠG_λ and component-wise multiplication [i.e., if $f, g \in \Pi G_\lambda$ then $(fg)\lambda = (f\lambda)(g\lambda)$ for all $\lambda \in \Lambda$]. Then each G_λ is identified with the subgroup $\{f \mid f_\mu = 1 \text{ for all } \mu \neq \lambda\}$. Prove that the

subgroup H generated by the union of these G_λ (in the direct product) coincides with the direct product if and only if $G_\lambda = \{1\}$ except for only a finite number of λ.

H constructed above is called the direct sum or the restricted direct product of these G_λ.

Note that the term direct sum often means direct product of a finite number of modules or rings.

EXERCISE 1.6

1. Let H and K be cyclic groups. Prove that if $H \times K$ is cyclic, either $\#(H) = 1$ or $\#(K) = 1$ or both $\#(H)$ and $\#(K)$ are finite and relatively prime to each other.
2. Give an elementary proof of Lemma 1.6.1.

CHAPTER 2

RINGS

In this chapter, we recall basic notions on rings, fields, and modules.

2.1 RINGS AND FIELDS

A *ring* (or, more precisely, an *associative ring*) is a set, say R, with two binary operations, called addition and multiplication, satisfying the following three conditions (1) to (3). As will be stated at the end of this section, we shall understand by a *ring* a commutative ring with identity except in this section.

(1) R is a commutative group under addition (associativity, commutativity, existence of zero, and existence of minus elements).

(2) R is a semigroup under multiplication (associativity).

(3) If $a, b, c \in R$, $\begin{cases} (a + b)c = ac + bc \\ c(a + b) = ca + cb \end{cases}$
(distributive property)

A ring R is called a *commutative* ring if the multiplication is commutative.

THEOREM 2.1.1 If R is a ring, then for arbitrary a, b R:

(i) $a0 = 0a = 0$

(ii) $(-a)b = a(-b) = -(ab)$, $(-a)(-b) = ab$

PROOF: Since $0 + 0 = 0$, we have $a(0 + 0) = a0$, and hence $a0 + a0 = a0$. Adding $-(a0)$, we have $a0 = 0$. Similarly, $0a = 0$. $(a + (-a))b = 0b = 0$, and hence $ab + (-a)b = 0$. Thus $(-a)b = -(ab)$. Similarly, $a(-b) = -(ab)$. $(-a)(-b) = -(a(-b)) = -(-(ab)) = ab$. QED

An element e of a ring R is called an identity if e is the identity of R under multiplication and if $e \neq 0$. If a ring R has an identity, then it is unique by Proposition 1.1.1 and is denoted by 1 or more explicitly by 1_R. An invertible element of R (under multiplication) is called a unit or an invertible element of R. Note that the set of units of R is a group, which is called the unit group of R. An element a of R is called a left zero-divisor if there is a nonzero element b such that $ab = 0$. Right zero-divisors are defined similarly. Right or left zero-divisors are called zero-divisors. An integral domain is a ring with identity in which every nonzero element is not a zero-divisor. An element a is called nilpotent if there is a natural number n such that $a^n = 0$. This a is a zero-divisor.

In general, if M is an additive group, for a natural number n and an element a of M, na denotes the n-ple sum $a + \cdots + a$ of a. Under the notation, we define the characteristic of a ring R as follows: Consider the set $C = \{n \in \mathbb{N} \mid na = 0 \text{ for every } a \in R\}$. If C is empty, then the characteristic of R is defined to be zero; otherwise, the least number in C is defined to be the characteristic of R. Note that if R has an identity, then $C = \{n \in \mathbb{N} \mid n1 = 0\}$.

If, in a ring R with identity, the unit group U coincides with $R - \{0\}$, then R is called a division ring or a field. Since our main aim is to study commutative fields, we shall understand by a field a commutative field from now on.

The set of all rational numbers, that of all real numbers, and that of all complex numbers are fields under the usual operations (addition and multiplication). They are called the rational number field, the real number field, and the complex number field and are denoted by $\mathbb{Q}$, $\mathbb{R}$, and $\mathbb{C}$, respectively. As for examples of rings, (1) the set of all rational integers is an integral domain, denoted by $\mathbb{Z}$, (2) the set of $n \times n$ matrices over a ring k is a ring, in which there are quite many zero-divisors and nilpotent elements [provided that $n \geq 2$ and that $\#(k)$ is not too small], and is called the full matric algebra of degree n over k.

If a subset S of a ring R forms a ring under the same operations, then S is called a subring of R. If furthermore S is a field, then S is called a subfield of R.

We add here a remark on identity elements. When a ring R is given, we can construct a new ring R* with identity as follows:

Considering the ring $\mathbb{Z}$ of rational integers, let R* be the product set $R \times \mathbb{Z}$, where we introduce addition and multiplication by $(r,a) + (s,b) = (r + s, a + b)$, $(r,a)(s,b) = (rs + sa + rb, ab)$. Then identifying each element r of R with $(r,0)$, R becomes a subring of R*, and $(0,1)$ is the identity of R*. Furthermore, because of the construction, many properties of R are carried over to R*. Therefore, when we want to know some properties of certain ring R without identity, we may look for the properties of R* and apply the results to R.

Thus, it is not a severe restriction to assume the existence of an identity in a ring.

As was stated at the beginning of this section, since we are to study commutative fields, we shall understand by a ring a commutative ring with identity, from now on, unless the contrary is explicitly stated.

2.2 HOMOMORPHISMS AND IDEALS

A homomorphism of a group was defined in Sec. 1.2 and is a mapping consistent with group operations. Similarly, a homomorphism of a ring is defined with respect to ring operations. Namely, a homomorphism ψ of a ring R into a ring R' is a mapping of R into R' such that $\psi(a + b) = \psi a + \psi b$ and $\psi(ab) = (\psi a)(\psi b)$. If K is a common subring of R and R' and if the restriction $\psi|_K$ of ψ to K is the identity mapping (i.e., $\psi x = x$ for every $x \in K$), then we say that ψ is a homomorphism over K or that ψ is a K-homomorphism.

For a homomorphism ψ of a ring R into a ring R', (1) the kernel of ψ is $\psi^{-1}(0)$; (2) if the kernel consists only of zero, then ψ is said to be injective or an injection; (3) if $\psi R = R'$, then we say that ψ is a homomorphism of R onto R' or that ψ is surjective or a surjection and (4) if ψ is injective and surjective, then we say that ψ is an isomorphism. An endomorphism of ring R is a homomorphism of R into itself, and an automorphism of R is an isomorphism of R to itself. The set of all homomorphisms (or K-homomorphisms) of R into R' is denoted by Hom (R,R') [or $\mathrm{Hom}_K(R,R')$, respectively]. The set of automorphisms (or K-automorphisms) of a ring R is a group and is denoted by Aut R (or Aut_K R, respectively). If, for rings R and R', there is an isomorphism (or K-isomorphism) of R to R', then we say that R is isomorphic (or K-isomorphic) to R' and we write $R \cong R'$ (or $R \cong_K R'$, respectively).

Let us observe the kernel I of a homomorphism ψ of a ring R into a ring R'. Since ψ is a homomorphisms of an additive group, I is an additive subgroup. Furthermore, I enjoys the property that

$$a \in I, \quad r \in R \quad \text{imply} \quad ar \in I$$

An additive subgroup I of a ring R enjoying this property is called an ideal of R. Then, as in the case of normal subgroups of groups, we have the following two theorems.

THEOREM 2.2.1 If I is an ideal of a ring R and if $I \neq R$, then R/I becomes a ring by defining multiplication by

$$(I + a)(I + b) = I + ab.$$

This ring R/I is called the residue class ring of R modulo I. On the other hand, R/R is defined to be $\{0\}$.

PROOF: As subsets of R we have $(I + a)(I + b) \subseteq I + ab$, and therefore the multiplication is well defined. Then the remaining part of the proof is straightforward. QED

THEOREM 2.2.2 If a homomorphism ψ of a ring R into a ring R' is given, (i) the kernel I of ψ is an ideal of R; (ii) the mapping $\psi b \rightsquigarrow I + b$ gives an isomorphism of ψR to R/I. Furthermore, under the correspondence $\psi R \supseteq S' \rightsquigarrow \psi^{-1}(S') = \{x \in R \mid \psi x \in S'\}$, we have (iii) one-one correspondence between the set of ideals of ψR and the set of ideals of R containing I.

PROOF: We know already that I is an ideal, and the remainder of the proof is similar to the case of groups. QED

For a given subset S of a group G, we defined the subgroup generated by S in Sec. 1.1. The definition is adapted to subrings and ideals. Namely, for a subset S of a ring R, (i) the smallest ideal containing S is the ideal generated by S and the ideal coincides with the intersection of all ideals containing S; and (ii) if S contains the identity, then the smallest subring containing S is well defined and is the intersection of all subrings containing S; the subring is called the subring generated by S.†

An ideal generated by a single element is called a principal ideal. If every ideal of a ring R is principal, then R is called a principal ideal ring; if furthermore R is an integral domain, then R is called a principal ideal domain.

Let I_λ ($\lambda \in \Lambda$) be ideals of a ring R. The sum of these I_λ is the ideal generated by the union of these I_λ and is denoted by $\sum I_\lambda$. The sum of ideals $I_1, \ldots, I_n$ may be written $I_1 + \cdots + I_n$. Note that $I_1 + \cdots + I_n = \{a_1 + \cdots + a_n \mid a_j \in I_j\}$. The product of a finite number of ideals $I_1, \ldots, I_n$ is defined to be the ideal generated by the set $\{a_1 \cdots a_n \mid a_j \in I_j\}$, and the product is denoted by $I_1 \cdots I_n$; in case $I = I_1 = \cdots = I_n$, the product defines the n-th power of the ideal I and is denoted by I^n.

2.3 DIRECT SUM OF RINGS

The direct sum of a finite number of rings $R_1, \ldots, R_n$ is defined similarly to the direct product of groups. Namely, it is the ring defined by component-wise operations on the set $R_1 \times \cdots \times R_n$, and the direct sum is denoted by $R_1 \oplus \cdots \oplus R_n$. Each R_i is identified with a subring of the direct sum, as in the case of groups. Then, denoting by e_i the identity of R_i, we see that $\sum_{i=1}^n e_i$ is the

† Since we are assuming the existence of identity in any ring, if S does not contain the identity of R, then there may not be the smallest subring containing S unless we define a subring so that it has the same identity as R.

identity of the direct sum, $e_ie_j = 0$ if $i \neq j$, and R_i is the ideal generated by e_i. As for the structural definition, it is given as follows: A ring R is the <u>direct sum</u> of its ideals $I_1, \ldots, I_n$ if R is the direct product (or direct sum; cf. the remark after Exercise 1.5.1) of $I_1, \ldots, I_n$ as additive groups. According to the group case, these direct sums may be called <u>direct products</u>.

We note here that, in the above structural definition, the assumption that I_j are ideals is important. Actually, under the assumption that each I_j is not zero, let $1 = \sum e_j$ with $e_j \in I_j$. Since $I_sI_t \subseteq I_s \cap I_t = 0$ if $s \neq t$, we see that $e_se_t = 0$ $(s \neq t)$. Therefore $1 = 1^2 = \sum e_j^{\,2}$ and we have $e_j^{\,2} = e_j$. $I_s = (\sum e_j)I_s = e_sI_s \subseteq e_sR \subseteq I_s$, and therefore $I_s = e_sR$, and I_s is a subring in which e_s is the identity, and the structure of R coincides with that of the direct sum of the rings $I_1, \ldots, I_n$ (under the constructive definition). If I_j were just subrings of R, then I_sI_t may not be contained in $I_s \cap I_t$, and we have a completely different situation.

THEOREM 2.3.1 (<u>Chinese Remainder Theorem</u>) Assume that $I_1, \ldots, I_s$ $(s \geq 2)$ are ideals of a ring R $(I_j \neq R$ for each $j)$ which enjoy the property that if $i \neq j$ then $I_i + I_j = R$. Then (1) if $i \neq j$, then $I_i \cap I_j = I_iI_j$; (2) if $\alpha_1, \ldots, \alpha_t$ and $\beta_1, \ldots, \beta_u$ are among $1, 2, \ldots, s$ and if there are no i and j such that $\alpha_i = \beta_j$, then $I_{\alpha_1}\cdots I_{\alpha_t} + I_{\beta_1}\cdots I_{\beta_u} = R$, hence, in particular, if $i \neq j$, then $I_i^{\,t} + I_j^{\,u} = R$; and (3) $R/(I_1\cap\cdots\cap I_s)$ is isomorphic to the direct sum of $R/I_1, \ldots, R/I_s$.

PROOF: (1): $I_iI_j \subseteq I_i \cap I_j = (I_i \cap I_j)R = (I_i \cap I_j)(I_i + I_j) \subseteq I_iI_j$. (2): $\Pi_j\ (I_{\alpha_j} + I_{\beta_\ell}) = R^t = R$, hence $I_{\alpha_1}\cdots I_{\alpha_t} + I_{\beta_\ell} = R$, and therefore $\Pi_\ell\ (I_{\alpha_1}\cdots I_{\alpha_t} + I_{\beta_\ell}) = R^u = R$; hence $I_{\alpha_1}\cdots I_{\alpha_t} + I_{\beta_1}\cdots I_{\beta_u} = R$. (3): For each $\alpha = 1, \ldots, s$, let ψ_α be the mapping $R/(I_1\cap\cdots\cap I_s) \to R/I_\alpha$ such that $(a \bmod I_1\cap\cdots\cap I_s) \rightsquigarrow (a \bmod I_\alpha)$. Each ψ_α is a surjective ring homomorphism. Let ψ be the mapping of $R/(I_1\cap\cdots\cap I_s)$ to the direct sum $D = (R/I_1) \oplus \cdots \oplus (R/I_s)$ such that $\psi x = \sum \psi_\alpha d$. Then ψ is a ring homomorphism. x is in the kernel of ψ if and only if x is in the kernel of all the ψ_α, which implies that $x = 0$. Thus ψ is an injection. In order to show that ψ is surjective, let $\sum a_i$ $(a_i \in R/I_i)$ be an arbitrary element of D. Since $I_1 + I_2 \cdots I_s = R$ by (2), there is a representative b_1 of a_1 such that $b_1 \in I_2 \cdots I_s$. Similarly, there is a representative b_i of a_i such that $b_i \in I_1 \cdots I_{i-1} I_{i+1} \cdots I_s$ for each $i = 2, \ldots, s$. Then we see easily that $\psi(b_1 + \cdots + b_s \mod I_1 \cap \cdots \cap I_s) = \sum a_i$. QED.

2.4 PRIME IDEALS

An ideal P of a ring R is called a prime ideal if R/P is an integral domain. Note that this condition implies that $P \neq R$.

THEOREM 2.4.1 For an ideal P of a ring R such that $P \neq R$, the following four conditions are equivalent to each other:

(1) P is a prime ideal of R.

(2) $a, b \in R$, $ab \in P$ imply that either a or b is in P.

(3) If A, B are ideals and if $AB \subseteq P$, then $A \subseteq P$ or $B \subseteq P$.

(4) If A, B are ideals such that $P \subset A$ and $P \subset B$, then $AB \nsubseteq P$.

PROOF: Obviously (1) implies (2). In order to prove that (2) implies (3), let us assume that $A \nsubseteq P$. Then there is $a \in A$ such that $a \notin P$. Then for each $b \in B$, $ab \in P$; hence $b \in P$ and $B \subseteq P$. Obviously (4) follows from (3). In order to complete the proof, let ψ be the natural homomorphism of R onto R/P, and assume that ψa $(a \in R)$ is a zero-divisor in R/P. Then there is $b \in R$ such that $\psi b \neq 0$ and $(\psi a)(\psi b) = 0$ (i.e., $ab \in P$). Set $A = aR + P$, $B = bR + P$. If $a \notin P$, then $A \supset P$, $B \supset P$ and we have $AB \nsubseteq P$ by (4), which contradicts the assumption that $ab \in P$. Thus $a \in P$ and R/P has no zero-divisor except zero. Thus (4) implies (1). QED

The following is important as an existence theorem of prime ideals.

THEOREM 2.4.2 Let S be a multiplicatively closed (i.e., $a, b \in S$ implies $ab \in S$) nonempty subset of a ring R. Assume that I is an ideal of R such that $I \cap S$ is empty. Let F be the set of ideals J of R which do not meet S. Then F has a maximal member (under the containment relation), and every maximal member of F is a prime ideal of R.

PROOF: It is easily seen that F is an inductive set, and therefore F has a maximal member, say P, by virtue of Zorn lemma. If ideals A and B contain P properly, then, by the maximality of P, $A \cap S$ and $B \cap S$ have elements s and t respectively. Then $st \in S$, $st \in AB$, and $AB \nsubseteq P$. Therefore P is prime by Theorem 2.4.1. QED

Under the notation as above, a maximal member of F is called a <u>maximal ideal with respect to</u> S. If S consists only of invertible elements, then a maximal member of F is called a <u>maximal ideal</u> of R.

COROLLARY 2.4.3 If A is an ideal of a ring R and if $A \neq R$, then there is a maximal ideal of R containing A. Every maximal ideal of R is a prime ideal.

PROPOSITION 2.4.4 (1) For an ideal M of a ring R, R/M is a field if and only if M is a maximal ideal. In particular: (2) R is a field (i) if and only if 0 is a maximal ideal, or (ii) if and only if R has no ideal except 0 and R.

PROOF: If R is a field and if A is an ideal $\neq 0$, then A has a nonzero element a and $1 \in aR \subseteq A$ (because $a^{-1} \in R$); hence $A \supseteq R$. Conversely, if R has no ideal except 0 and R, then for each nonzero element a of R, the ideal aR coincides with R; hence a has its inverse in R. Thus R is a field. This proves (2), (ii). The equivalence of (i) and (ii) in (2) is obvious, and we have proved (2). (1) follows from (2) by virtue of Theorem 2.2.2. QED

The following lemma is useful.

LEMMA 2.4.5 If $P_1, \ldots, P_n$ are prime ideals of a ring R and if A is an ideal of R such that $A \nsubseteq P_i$ for every i, then there is an element a of A which is not in any P_i.

PROOF: We employ an induction on n. If $n = 1$, then the assertion is obvious and we assume that $n \geq 2$. Let b be an element of A which is not in any of $P_1, \ldots, P_{n-1}$. If $b \notin P_n$, then we set $a = b$. Assume that $b \in P_n$. Since we may assume that $P_i \not\subseteq P_n$ $(i < n)$, $AP_1 \cdots P_{n-1} \not\subseteq P_n$ and there is $c \in AP_1 \cdots P_{n-1}$ such that $c \notin P_n$. Then $a = b + c$ is the required element. QED

Let D be an integral domain. A field K is called a <u>field of fractions</u> (or a <u>field of quotients</u>) of D if (i) D is a subring of K and (ii) every element c of K is expressed in the form ab^{-1} with $a,b \in D$. Under the definition, we have:

THEOREM 2.4.6 Every integral domain D has its field K of fractions, which is unique within isomorphisms; i.e., if K^* is another field of fractions of D, then K^* is D-isomorphic to K.

PROOF: In the set $Q = \{(a,b) \mid a,b \in D,\ b \neq 0\}$, we introduce a relation $\sim$ by $(a,b) \sim (a',b')$ if and only if $ab' = a'b$. Then this is an equivalence relation as is easily seen. The class of (a,b) is denoted by a/b, and in the set K of equivalence classes, we introduce addition and multiplication by

$$(a/b) + (a'/b') = (ab' + a'b)/(bb'), \qquad (a/b)(a'/b') = (aa')/(bb')$$

Then, as is easily seen, K is a field whose identity is $1/1$ and zero is $0/1$. Furthermore, one can identify a in D with $a/1$ in K, and D can be regarded as a subring of K. Then a/b coincides with ab^{-1}. Thus K is a field of fractions of D. Assume that K^* is another field of fractions of D. To each a/b in K, let us correspond ab^{-1} in K^*. $a/b = a'/b'$ if and only if $ab' = a'b$ which is equivalent to $ab^{-1} = a'b'^{-1}$. Therefore the correspondence is a one-one correspondence between K and K^*. Since $ab^{-1} + a'b'^{-1} = (ab' + a'b)(bb')^{-1}$ and $(ab^{-1})(a'b'^{-1}) = aa'(bb')^{-1}$, we see that the correspondence gives an isomorphism. QED

This method of constructing the field of fractions is adapted to a general ring. Namely, let R be a ring and let S be the set of non-zero-divisors in R. In the set $Q = \{(a,s) \mid a \in R,\ s \in S\}$, we introduce an equivalence relation as above and construct a ring K containing R which enjoys the properties that (1) every element of S is a unit in K and (2) every element of K is expressed in the form as^{-1} with $a \in R$, $s \in S$. Such a ring K is called a ring of total fractions, or total quotient ring of R. Uniqueness is proved quite similarly.

2.5 POLYNOMIAL RINGS

Let R be a ring and let $X_1, \ldots, X_n$ be n letters (i.e., symbols). A form of the type $X_1^{i_1} \cdots X_n^{i_n}$ (each i_j being a nonnegative rational integer) [sometimes $cX_1^{i_1} \cdots X_n^{i_n}$ $(c \in R)$, too] is called a monomial in variables $X_1, \ldots, X_n$, and $\sum_j i_j$ is called the degree of the monomial. Multiplication of monomials is defined by

$$(X_1^{i_1} \cdots X_n^{i_n})(X_1^{j_1} \cdots X_n^{j_n}) = X_1^{i_1+j_1} \cdots X_n^{i_n+j_n}$$

A linear combination $\sum_{i=1}^{s} a_i m_i$ of monomials $m_1, \ldots, m_s$ with coefficients $a_1, \ldots, a_s$ in R is called a polynomial in variables $X_1, \ldots, X_n$ over (or with coefficients in) R. Terms with zero coefficients are disregarded; namely, (1) if a monomial m does not appear in a polynomial f, then 0 is the coefficient of m in f, and (2) two polynomials f and g are equal to each other if and only if, for each monomial m, the coefficients of m in f and g coincide to each other. For a nonzero polynomial f, the maximum of the degree of the monomials with nonzero coefficients in f is called the degree of f and is denoted by $\deg f$. The degree of zero is indefinite, i.e., every rational integer can be $\deg 0$. On the other hand, if a polynomial f is expressed in the form $\sum a_i m_i$ with monomials m_i and coefficients a_i, and if d is the maximum of the degrees of m_i, then we say that f is of degree d in appearence. A polynomial $\sum a_i m_i$ $(a_i \in R)$, such that all of these minomials m_i are of the same degree, is called a homogeneous form.

A polynomial f in one variable X over R is expressed in the form $c_0X^d + c_1X^{d-1} + \cdots + c_d$ $(c_i \in R)$. If $c_0 = 1$, then f is said to be monic.

Addition and multiplication of polynomials are defined by

$$(\textstyle\sum a_i m_i) + (\sum b_i m_i) = \sum (a_i + b_i) m_i$$

$$(\textstyle\sum a_i m_i)(\sum b_i m_i) = \sum_{m_i} (\sum_{m_\alpha m_\beta = m_i} a_\alpha b_\beta) m_i$$

Then, as is easily seen, the set of polynomials in $X_1, \ldots, X_n$ over

R forms a ring. This new ring is called the <u>polynomial ring</u> in the variables $X_1, \ldots, X_n$ over R and is denoted by $R[X_1,\ldots,X_n]$. Note that $R[X_1,\ldots,X_n] = R[X_1,\ldots,X_{n-1}][X_n]$. As for this notation, we must note that, in general, if R is a subring of another ring R' with common identity and if R' is generated by elements a_1, $\ldots$, a_n over R, then R' is expressed by $R[a_1,\ldots,a_n]$. In this case, there is an R-homomorphism ψ of the polynomial ring $R[X_1,\ldots,X_n]$ onto $R[a_1,\ldots,a_n]$ such that $\psi X_i = a_i$ for every i. For each $f(X_1,\ldots,X_n) \in R[X_1,\ldots,X_n]$, $f(a_1,\ldots,a_n)$ denotes the element $\psi(f(X_1,\ldots,X_n))$. Namely, if $f(X_1,\ldots,X_n) = \sum c_{i_1\cdots i_n} X_1^{i_1}\cdots X_n^{i_n}$ then $f(a_1,\ldots,a_n) = \sum c_{i_1\cdots i_n} a_1^{i_1}\cdots a_n^{i_n}$. We say that $a_1, \ldots, a_n$ are <u>algebraically independent</u> over R if the kernel of ψ is the zero ideal; otherwise, we say that $a_1, \ldots, a_n$ are <u>algebraically dependent</u> over R. The kernel of ψ is called the <u>relation ideal</u> for $a_1, \ldots, a_n$. If $a_1, \ldots, a_n$ are algebraically independent over R, we may say that $R[a_1,\ldots,a_n]$ is a <u>polynomial ring</u>.

PROPOSITION 2.5.1 If D is an integral domain, then the polynomial ring $D[X_1,\ldots,X_n]$ is also an integral domain.

PROOF: It suffices to prove the case $n = 1$, $X_1 = X$. Nonzero polynomials f, g are expressed as $f = a_0X^d + \cdots + a_d$, $g = b_0X^e + \cdots + b_e$ ($a_i, b_j \in D$, $a_0 \neq 0$, $b_0 \neq 0$). Then the coefficient of X^{d+e} in fg is equal to a_0b_0 and we see that $fg \neq 0$. QED

COROLLARY 2.5.2 If f, g are nonzero polynomials over an integral domain D, then $\deg fg = \deg f + \deg g$.

PROOF: $f = f_0 + \cdots + f_d$, $g = g_0 + \cdots + g_e$ where f_i and g_j are homogeneous forms of degrees i and j, respectively, $d = \deg f$, $e = \deg g$. Then fg is the sum of all f_ig_j and $f_dg_e \neq 0$. Hence we see the assertion. QED

COROLLARY 2.5.3 If P is a prime ideal of a ring R, then the ideal $PR[X_1,\ldots,X_n]$ of the polynomial ring is a prime ideal.

PROOF: The natural homomorphism $\psi : R \to R/P$ extends naturally to a homomorphism $\psi' : R[X_1,\ldots,X_n] \to (R/P)[X_1,\ldots,X_n]$, and the kernel of ψ' is $PR[X_1,\ldots,X_n]$. QED

THEOREM 2.5.4 In the polynomial ring $R[X]$ in one variable X over a ring R, we have:

(1) If h is a monic polynomial, then (i) for each $f \in R[X]$, $\deg fh = \deg f + \deg h$; (ii) h is not a zero-divisor; and (iii) for each $f \in R[X]$; there are $q,r \in R[X]$ such that $f = qh + r$, $\deg r < \deg h$, and such a pair (q,r) is unique.

This r is called the remainder of f modulo h. If $r = 0$, then we say that h divides f or that h is a factor of f or that f is divisible by h (cf. Sec. 2.6).

(2) If $f(X) \in R[X]$ and if $a \in R$, then the remainder of $f(X)$ modulo $X - a$ coincides with $f(a)$ (remainder theorem). In particular, if $f(a) = 0$, then $X - a$ divides $f(X)$ (factor theorem).

In this last case, we call a a root of $f(X)$ [or of the equation $f(X) = 0$].

PROOF: (1), (i) is proved similarly as in Proposition 2.5.1, and (ii) follows from (i). As for (iii), let $d = \deg h$ and $e = \deg f$. If $e < d$, then $q = 0$, $r = f$ enjoy the property. So we assume that $e \geq d$, and we let c be the coefficient of X^e in f. Then, observing $f - cX^{e-d}h$ instead of f, we prove the existence of (q,r) by induction on e. Assume now that another pair (q',r') enjoys the property for (q,r). Then $0 = (q - q')h + (r - r')$. If $q - q' \neq 0$, then by (i), $\deg (r - r') = d + \deg (q - q') \geq d$ which contradicts that $\deg r < d$, $\deg r' < d$.

(2): By (1), $f(X) = q(X)(X - a) + r$, $\deg r < \deg (X - a) = 1$, and therefore $r \in R$. Then $f(a) = q(a)(a - a) + r = r$. QED

COROLLARY 2.5.5 In the above theorem, assume that R is an integral domain and that, for $f(X)$ ($\in R[X]$), there are mutually distinct roots $a_1, \ldots, a_n$ in R. Then $f(X)$ is divisible by $(X - a_1)\cdots(X - a_n)$. Hence, in particular, if furthermore $\deg f(X) < n$, then $f(X) = 0$.

PROOF: We employ induction on n, and we may assume that $f(X) = (X - a_1)\cdots(X - a_{n-1})g(X)$, and we have $0 = f(a_n) = (a_n - a_1)\cdots(a_n - a_{n-1})g(a_n)$. Since R is an integral domain, we have $g(a_n) = 0$ and $X - a_n$ divides $g(X)$. QED

2.6 UNIQUENESS OF FACTORIZATIONS

Let R be a ring. An element p of R is called a prime element if pR is a prime ideal $\neq 0$. For two elements a and b of R, if there is an element c of R such that $a = bc$, then we say that a is a multiple of b, or that b is a divisor or factor of a, or that a is divisible by b, or b divides a, in R. (Note that divisibility depends on the ring we are concerned with.) If furthermore c can be a unit, then we say that a and b are associates of each other. Then, for two elements a, a' of R, we can consider a common divisor (common factor) c which is a divisor of both a and a'. If there is a common divisor c which is divided by every common divisor, then c is called the greatest common measure or the largest common factor. Similarly, we define common multiples and the least common multiple.

An element a of R is said to be irreducible if divisors of a are units and associates of a only. Under this definition, a unit is irreducible, and therefore an irreducible element which is not a unit is called a proper irreducible element.

THEOREM 2.6.1 Every prime element p of an integral domain D is irreducible.

PROOF: If $p = bc$ $(b,c \in D)$, then either b or c is in pD. If, for instance, $b \in pD$, then $b = pr$ $(r \in D)$, and $p = prc$. Since D is an integral domain and since $p \neq 0$, we have $1 = rc$. QED

An integral domain D is called a unique factorization domain, or simply a UFD, if every nonzero nonunit is expressed as a product of a finite number of prime elements. This terminology is justified by:

THEOREM 2.6.2 An integral domain D is a UFD if and only if the following two conditions are satisfied, and in that case, every proper irreducible element in D is a prime element.

(i) Every nonzero element is a product of a finite number of irreducible elements.

(ii) If $b_1, \ldots, b_m, c_1, \ldots, c_n$ are proper irreducible elements and if $b_1 \cdots b_m = c_1 \cdots c_n$, then $m = n$ and $b_i = c_{\sigma i} u_i$ $(i = 1, \ldots, n)$ with units u_i and a suitable permutation σ on $\{1,..,n\}$,

PROOF: Assume first that D is a UFD. Then (i) follows from Theorem 2.6.1. We prove (ii) by induction on n. If $n = 1$, then $b_1 \cdots b_m$ is irreducible; hence $m = 1$. Assume that $n > 1$. By our assumption c_n is a product of prime elements and is irreducible; hence c_n is a prime element. Therefore some b_i, say b_m, must be divisible by c_n. Since b_m is irreducible, b_m must be an associate of c_n. Thus $c_n u = b_m$ with a unit u. Then $b_1 \cdots b_{m-1} u = c_1 \cdots c_{n-1}$, and our induction assumption proves (ii). Conversely, assume that (i) and (ii) hold in D, and assume that p is a proper irreducible element in D. If $bc \in pD$ $(b,c \in D)$, then $bc = pq$ $(q \in D)$. Express b, c, and q as products of irreducible elements: $b = b_1 \cdots b_m$, $c = c_1 \cdots c_n$, $q = q_1 \cdots q_s$ and $b_1 \cdots b_m c_1 \cdots c_n = pq_1 \cdots q_s$. By (ii), there must be an associate of p among $b_1, \ldots, b_m, c_1, \ldots, c_n$, which shows that b or c is in pD. Thus every proper irreducible element in D is a prime element and D is a UFD. QED

There are some other characterizations for UFD as in Exercise 2.6.1. An integral domain E is called a <u>Euclid ring</u> if there is a mapping v of E into a well-ordered set W so that the following two conditions are satisfied:

(i) If $0 \neq a \in E$, then $v0 < va$.

(ii) If $a,b \in E$, $a \neq 0$, then there are $q,r \in E$ so that $b = aq + r$, $vr < va$.

The ring of rational integers is a Euclid ring (va being the absolute value of a). The polynomial ring $K[X]$ of one variable X over a field K is also a Euclid ring (va being the degree of a if $a \neq 0$).

THEOREM 2.6.3 (1) Every Euclid ring E is a principal ideal domain.

(2) Every principal ideal domain D is a UFD, and if P is a prime ideal of D, then P is either a maximal ideal or zero.

PROOF: (1): Let A be an arbitrary ideal of E, and we shall show that A is principal. We may assume that $A \neq 0$. Let v be as in the definition of a Euclid ring and set $J = \{va \mid 0 \neq a \in A\}$. J is a nonempty subset of a well-ordered set; hence J has the smallest member, say vm $(m \in A)$. For an arbitrary element b of A, there are $q, r \in E$ so that $b = mq + r$, $vr < vm$. Since $r \in A$, our choice of m implies that r must be zero. Thus $A \subseteq mE$, and hence $A = mE$. Thus (1) is proved.

(2): Let us prove that if a $(\neq 0)$ is a nonunit in D, then a is a product of prime elements. There is a maximal ideal M_1 of D which contains a. $M_1 = p_1D$ with a prime element p_1 and $a = a_1p_1$. If a_1 is a nonunit, we may apply the same to a_1, and so on, and we obtain a sequence of factorizations $a = p_1 \cdots p_n a_n$ with prime elements $p_1, \ldots, p_n$. If some a_n is a unit, then a is a product of prime elements, as we wished. Assume the contrary. Then we have an infinite ascending sequence $aD \subset a_1D \subset \cdots \subset a_nD \subset \cdots$. Then $B = \cup_{i=1}^{\infty} a_iD$ is an ideal of D, and hence $B = bD$ $(b \in B)$. Then $b \in a_nD$ with an n, and $B \subseteq a_nD$, which is a contradiction. Thus D is a UFD. Now, if the prime ideal P is not zero, then $P = pD$ $(p \neq 0)$. Then P is contained in a maximal ideal $M = qD$, and p is divisible by q. Since p is irreducible by Theorem 2.6.1, we see that $P = M$. QED

In the remaining part of this section, we want to prove the following:

THEOREM 2.6.4 If R is a UFD, then the polynomial ring $R[X_1, \ldots, X_n]$ of n variables $X_1, \ldots, X_n$ over R is also a UFD. In particular, the polynomial ring in n variables over a field is a UFD.

PROOF: By virtue of an induction argument, it suffices to prove the case where $n = 1$, $X = X_1$. We introduce here the notion of a <u>primitive</u> polynomial as a polynomial $a_0X^n + \cdots + a_n$ whose coefficients have no nontrivial common divisor (except units). We note here that, by virtue of Corollary 2.5.3,

(*) <u>every prime element of R is a prime element in $R[X]$</u>.

Now we prove a lemma:

LEMMA 2.6.5 Let K be the field of fractions of R. Let $f, g \in R[X]$ and assume that g is primitive. If g divides f in $K[X]$, then g divides f in $R[X]$.

PROOF: Set $h = f/g$ and let d be a nonzero element of R such that $dh \in R[X]$. Then we have $dh = eh'$ with $e \in R$ and a primitive polynomial h'. We may assume that d and e have no non-trivial common divisor. Then $df = dhg = eh'g$. If d is divisible by a prime element p of R, then by (*) above, one of e, h', g must be divisible by p. Since h' and g are primitive and since e has no common divisors with d, d has no prime factor which implies that d is a unit; hence $h \in R[X]$. QED

COROLLARY 2.6.6 If a primitive polynomial g in $R[X]$ is a prime element in $K[X]$, then g is a prime element in $R[X]$.

PROOF: If $f, h \in R[X]$ and $fh \in gR[X]$, then f or h are divisible by g in $K[X]$, and hence in $R[X]$, too, by virtue of the lemma. QED

Now we come back to the proof of Theorem 2.6.4. Take a nonzero polynomial f ($\in R[X]$). Then f is expressed as ah with $a \in R$ and a primitive polynomial h. By (*) above, a is expressed as a product of prime elements. As for h, h is a product of prime elements in $K[X]$, because $K[X]$ is a Euclid ring. A prime factor p_1 of h in $K[X]$ can be assumed to be a primitive polynomial in $R[X]$, and $h = p_1h_1$ with $h_1 \in R[X]$ by virtue of Lemma 2.6.5. Hence, using an induction on the degree of h, we see that h is a product of prime elements in $R[X]$ by virtue of Corollary 2.6.6. QED

We add here some remarks on Euclid rings. Let us employ the notation in the definition (just before Theorem 2.6.3). Since $b = aq + r$, we see that an element c of E is a common divisor of a and b if and only if it is a common divisor of b and r. Therefore if two elements a, b of E are given, then the largest common factor c of a, b exists and is computed by the so-called Euclid algorithm, which is as follows: Start with $a = a_0$, $b = a_1$, and define a_{i+1} inductively by $a_{i-1} = a_i q + a_{i+1}$ with $q \in E$ and $va_{i+1} < va_i$. This is done until some a_i, say a_{m+1}, become zero. Then the largest common factor of a_i and a_{i+1} does not depend on i, whence a_m is the largest common factor of a and b.

This remark and the fact that the polynomial ring K[X] is a Euclid ring if K is a field give us the following result:

PROPOSITION 2.6.7 Let K be a subfield of a field L and let f(X), g(X) be polynomials in X over K. If f(X) and g(X) have a common root as polynomials over L, then they have a nontrivial common factor as polynomials over K.

PROOF: The computation of the largest common factor given above is independent of whether "over K" or "over L." The assumption implies that the largest common factor is nontrivial. QED

By the way, we define the multiplicity of a root. Let f(X) be a polynomial in X over an integral domain R and let a be an element of R. Then there is a natural number m such that f(X) is divisible by $(X - a)^{m-1}$ but not by $(X - a)^m$. Then m - 1 is called the multiplicity of a as a root of f(X). [a is a root of f(X) if and only if m - 1 > 0, and in that case we say that a is an (m - 1)-ple root of f(X).] A root of f(X) is called a simple root if its multiplicity is one; otherwise it is called a multiple root. The fact that K[X] is a UFD for the field K of fractions of R shows that the multiplicity is well defined and that the sum of all multiplicities of roots of f(X) in K is at most the degree of f(X).

2.7 MODULES

Let R be a ring. M is called a left R-module if (1) M is a module (i.e., a commutative additive group) and (2) an operation, called a multiplication from the left, of elements of R on M is defined (i.e., for $r \in R$ and $m \in M$, the product rm is well defined and $rm \in M$) so that the following three conditions are satisfied:

(i) If $r \in R$ and $m, m' \in M$, then $r(m + m') = rm + rm'$.

(ii) If $r, r' \in R$ and $m \in M$, then $r(r'm) = (rr')m$, $(r + r')m = rm + r'm$.

(iii) As for the identity of R, it holds $1m = m$ for every $m \in M$.

A right R-module is defined similarly, with reverse order of multiplication. But, because of the commutativity of R, if M is a right R-module or a left R-module, then defining $rm = mr$ for every $(r,m) \in R \times M$, M becomes a left or right R-module, too. Therefore, in this text, we do not distinguish "left" or "right," and we called them R-modules or modules over R. [In some of the literature, our R-modules are called unitary R-modules, and the notion of an R-module is defined without condition (iii) above.]

When M is an R-module, we understand by a submodule of M a subset of M which is an R-module under the same operations; if we want to emphasize that we are dealing with R-modules, we call it an R-submodule. Note that an ideal of R is nothing but a submodule of R, R being an R-module.

Let M and N be modules over R. An R-homomorphism f of M into N is a homomorphism of M into N as additive groups with an additional condition that $r(fm) = f(rm)$ for every $(r,m) \in R \times M$. An R-homomorphism may be called a homomorphism of modules over R. Similar definitions are given to the case of surjections, injections, and isomorphisms. Then the results in Sec. 1.2 are adapted to our case. For instance, Theorems 1.2.1 to 1.2.3 become as follows:

THEOREM 2.7.1 (1) If N is a submodule of a module M over a ring R, then the natural homomorphism of M onto M/N is an R-homomorphism.

(2) If ψ is an R-homomorphism of an R-module M into an R-module M'. then (i) the kernel N of ψ is an R-submodule of M and $\psi M \cong M/N$, and (ii) there is a one-one correspondence between the set of submodules of ψM and the set of submodules of M containing N so that H' corresponds to H if and only if $\psi H = H'$.

(3) If N and N' are submodules of an R-module M, then the natural homomorphism of M onto M/N induces an isomorphism of

$N'/(N \cap N')$ to $(N + N')/N$ (isomorphism theorem).

For the detail of the proof, we leave it to the readers.

An R-module (R being a ring) M is generated by a nonempty subset S if and only if $M = \sum_{s \in S} sR$, i.e., M is the smallest R-module containing S. Therefore we define $\{0\}$ as the submodule generated by the empty set. M is said to be finitely generated or of finite type or simply finite if such S can be chosen to be a finite set.

Elements $a_1, \ldots, a_m$ of an R-module M are said to be linearly independent over R if $\sum_{i=1}^{m} r_i a_i = 0$ with $r_i \in R$ implies that all r_i are zero. A subset S of M is said to be linearly independent if every nonempty finite subset of S consists of linearly independent elements. Hence, we understand that the empty set is linearly independent.

If an R-module M is generated by a linearly independent subset S, then we say that M is a free R-module and that S is a free base or a linearly independent base for M.

THEOREM 2.7.2 Every module M over a field K is free.

PROOF: Let F be the set of linearly independent subsets of M. Then F is an inductive set. Indeed, (i) the empty set belongs to F, and hence F is not empty; and (ii) if a subset $\{S_\lambda \mid \lambda \in \Lambda\}$ of F is well-ordered, and if T is a finite subset of $S^* = \cup_\lambda S_\lambda$, then there is an S_μ which contains T and T consists of linearly independent element. This implies that $S^* \in F$, and $S^* = \sup \{S_\lambda\}$. Therefore, by Zorn lemma, there is a maximal member B of F. It suffices to show that the submodule N generated by B coincides with M. Assume the contrary, and let a be an element of M outside of N. By the maximality of B, we see that $B \cup \{a\}$ is not linearly independent. This implies that there are elements $b_1, \ldots, b_m$ of B such that $a, b_1, \ldots, b_m$ are linearly dependent; hence there is a nontrivial relation $c_0 a + \sum_i c_i b_i = 0$ $(c_i \in K)$. Since $b_1, \ldots, b_m$ are linearly independent, we see that $c_0 \neq 0$. Since K is a field, it contains c_0^{-1}; hence we may assume that $c_0 = 1$, and therefore $a = -(\sum c_i b_i) \in N$, which contradicts our assumption. QED

From the proof above, we note the following:

COROLLARY 2.7.3 If M is a module over a field K and if S is a linearly independent subset of M, then there is a free base B for M such that $S \subseteq B$.

When M and N are modules over a ring R, the set of R-homomorphisms of M into N is denoted by $\mathrm{Hom}_R(M,N)$ or simply by $\mathrm{Hom}(M,N)$ if it is obvious that we are observing R-homomorphisms only. In $\mathrm{Hom}(M,N)$, we introduce addition by $(f_1 + f_2)m = f_1m + f_2m$ ($f_i \in \mathrm{Hom}(M,N)$, $m \in M$), and we see easily that $f_1 + f_2 \in \mathrm{Hom}(M,N)$. The zero element of $\mathrm{Hom}(M,N)$ is the homomorphism f such that $fM = 0$, and this f is denoted by 0 usually. Furthermore, a multiplication of elements of R on $\mathrm{Hom}(M, N)$ is defined by $(rf)m = r(fm)$ ($f \in R$, $f \in \mathrm{Hom}(M,N)$, $m \in M$). Thus:

PROPOSITION 2.7.4 $\mathrm{Hom}(M,N)$ is an R-module.

Let us be concerned with the notion of the length of a module. For an ascending or a descending chain of sets, its <u>length</u> is defined to be the number of intervals; for instance, the length of $M = M_0 \supset M_1 \supset \cdots \supset M_r$ is r. To <u>refine</u> such a chain means to make a longer chain inserting suitable terms in intervals; for instance, $M_0 \supset M_1 \supset N \supset M_2 \supset \cdots \supset M_r$ is a refinement of $M_0 \supset M_1 \supset M_2 \supset \cdots \supset M_r$. For a given module M over a ring R, a descending chain of submodules of M, say $M_0 \supset M_1 \supset \cdots \supset M_r$, is called a <u>composition series</u> of M if (i) $M_0 = M$, $M_r = 0$ and (ii) there is no refinement of this chain among descending chains of submodules of M.

The notion of composition series and the following theorem are really handled in group theory. But we observe the module case only in this text because (1) we need only the case of modules and (2) if the readers wish they can adapt our proof to the case of groups easily.

THEOREM 2.7.5 (<u>Jordan-Hölder-Schreier Theorem</u>) Assume that a module M over a ring has a composition series $M = M_0 \supset M_1 \supset \cdots \supset M_n = 0$. If $M = N_0 \supset N_1 \supset \cdots \supset N_s = 0$ is a descending chain of submodules, then (i) $s \leq n$ and (ii) $s = n$ if and only if $N_0 \supset \cdots \supset N_s$ is a composition series. In the latter case, there is a permutation σ on $\{1,2,\ldots,n\}$ so that $N_{i-1}/N_i \cong M_{\sigma i-1}/M_{\sigma i}$.

This number n is called the <u>length</u> of the R-module M and is denoted by $\text{length}_R M$, or length M. (If a module M has no composition series, we define the <u>length</u> of M to be infinite.) If R is a field, this may be called the <u>dimension</u> of M viewing M as a vector space over R (see Theorem 2.7.6).

PROOF: We use induction on n. If $n \leq 1$, then the assertion is obvious, and we assume that $n \geq 2$.

(1) Assume that $N_1 \subseteq M_1$. In this case, since M_1 has a composition series of length $n - 1$ and since we may assume that $N_1 = M_1$ (inserting M_1 between M and N_1), we prove the assertion by induction hypothesis applied to M_1.

(2) Assume that $N_1 \not\subseteq M_1$ and $N_1 \cap M_1 = 0$. In this case, since there is no proper submodule properly containing M_1, we have $M = N_1 + M_1$; hence $M = M_1 \oplus N_1$, $M/M_1 = N_1$, and $M/N_1 = M_1$. Therefore $s = 2 \leq n$; assertion (ii) and the last assertion follow from the induction hypothesis applied to M_1.

(3) Assume now that $N_1 \not\subseteq M_1$ and $N_1 \cap M_1 = D \neq 0$. By the induction hypothesis applied to M_1, there is a composition series $M = A_0 \supset M_1 = A_1 \supset \cdots \supset A_t = D \supset A_{t+1} \supset \cdots \supset A_n = 0$, together with a permutation τ on $\{1,2,\ldots,n\}$ such that $\tau 1 = 1$ and $A_{i-1}/A_i \cong M_{\tau i-1}/M_{\tau i}$ for each $i = 1, \ldots, n$. Then M/D has a composition series of length t, and $t < n$ because $D \neq 0$. Therefore, there is a composition series of M/D going through N_1/D and of length t; namely, there is a composition series of M such that $M = B_0 \supset B_1 \supset \cdots \supset B_c = N_1 \supset B_{c+1} \supset \cdots \supset B_t = D = A_t \supset \cdots \supset B_n = A_n = 0$, together with a permutation η on $\{1,\ldots,n\}$ such that $B_{i-1}/B_i \cong A_{\eta i-1}/A_{\eta i}$ ($\eta j = j$ for $j \geq t$), whence $B_{i-1}/B_i \cong M_{\tau\eta i-1}/M_{\tau\eta i}$ for each $i = 1, \ldots, n$. Then we may employ $B_0 \supset \cdots \supset B_n$ instead of $M_0 \supset \cdots \supset M_n$, and we have reduced to the case (1). QED

As for modules over a field, the theorem above can be stated as follows:

THEOREM 2.7.6 Assume that a module M over a field K has a free base $b_1, \ldots, b_n$ and that $c_1, \ldots, c_s$ are linearly independent elements of M. Then (i) $s \le n$, (ii) $s = n$ if and only if $c_1, \ldots, c_s$ form a free base for M, and (iii) there is a permutation σ on $\{1, \ldots, n\}$ such that $c_1, \ldots, c_t, b_{\sigma(t+1)}, \ldots, b_{\sigma n}$ form a free base for M for every $t = 1, .., s$.

Instead of the proof, we give remarks on the relationship between this and the former theorem and on the proof. In the above case, $b_1, \ldots, b_n$ form a free base if and only if $M_0 = M \supset M_1 = \sum_{j>1} b_jK \supset \cdots \supset M_i = \sum_{j>i} b_jK \supset \cdots \supset M_n = 0$ form a composition series. Hence (i) and (ii) correspond to (i) and (ii) in the previous theorem, respectively. As for the proof, we can give a simpler proof. Namely, first we show (iii) with modification so that the last part is "for each $t = 1, 2, \ldots, \min\{s, n\}$." This is done easily by induction on t (and we leave it to the readers). Then (i) and (ii) follow immediately.

A module M over a ring R is called an <u>Artin module</u> if it has a composition series.† If the ring R itself is an Artin R-module, then we say that R is an <u>Artin ring</u>. We prove here some easy facts on Artin rings and Artin modules (and some more will be proved in Chap. 4).

THEOREM 2.7.7 Let M be an Artin module over a ring R.

(1) $\text{length}_R M = 1$ if and only if $M \cong R/P$ with a maximal ideal P.

(2) Assume that R contains a field K, that $\text{length}_K R/P$ is finite for all maximal ideals P of R, and that the value s of $\text{length}_K R/P$ is independent of the choice of P. Then we have $s(\text{length}_R M) = \text{length}_K M$.

PROOF: (1): The "if part" is obvious and we assume that $\text{length}\, M = 1$. Let $0 \neq a \in M$ and consider the mapping f of R into M defined by $fx = xa$. Then fR is a submodule of M different from zero. Hence $fR = M$ and $M \cong R/(f^{-1}(0))$; $f^{-1}(0)$ is maximal.

---

† In some of the literature, an Artin module is defined to be a module satisfying the minimum condition for submodules. An Artin module in this sense may not be an Artin module in our sense.

(2): Let $M = M_0 \supset M_1 \supset \cdots \supset M_n = 0$ be a composition series of M. Each M_{i-1}/M_i is isomorphic to R/P_i with a maximal ideal P_i. By our assumption, $\text{length}_K R/P_i = s$, which means that each chain $M_{i-1} \supset M_i$ can be refined by K-submodules and we obtain a descending chain of length s which corresponds to a composition series of R/P_i. Thus we obtain a composition series of the K-module M of length sn. QED

THEOREM 2.7.8 If R is an Artin ring, then (i) R has only a finite number of prime ideals and (ii) all prime ideals are maximal.

PROOF: If a prime ideal P is not maximal, then there is $a \in R$ such that $a' = (a \bmod P)$ is a nonzero nonunit in $R' = R/P$. Then $a'^n R' \supset a'^{n+1} R'$. Indeed, otherwise, $a'^n = a'^{n+1} b'$ with $b' \in R'$ and $a'b' = 1$ (because R/P is an integral domain and $a' \neq 0$), which contradicts our assumption that a' is a nonunit. Therefore we have an infinite descending chain of ideals $I_1 \supset I_2 \supset \cdots$ with $I_n = a^n R + P$, which contradicts our assumption that R is an Artin ring. Thus we have proved (ii). Assume now that there are infinitely many maximal ideals $P_1, \ldots, P_n, \ldots$. Then by Lemma 2.4.5 we have an infinite descending chain of ideals $P_1 \supset P_1 \cap P_2 \supset \cdots \supset P_1 \cap \cdots \cap P_n \supset \cdots$ which is a contradiction. QED

2.8 SYMMETRIC FORMS AND ALTERNATING FORMS

Consider the polynomial ring $P = R[X_1, \ldots, X_n]$ in n variables over a ring R. Then the symmetric group S_n on $\{1, \ldots, n\}$ acts on P by the multiplication defined by:

$$f(X_1, \ldots, X_n) = f(X_{\sigma 1}, \ldots, X_{\sigma n}) \qquad (\sigma \in S_n)$$

An element f of P is said to be symmetric if $\sigma f = f$ for every $\sigma \in S_n$. f is said to be alternating if (i) f is not symmetric and (ii) $\sigma f = \pm f$ for every $\sigma \in S_n$.

When $a_1, \ldots, a_n$ are elements of a ring containing R, $f(a_1, \ldots, a_n)$ is called a symmetric form, or an alternating form, on $a_1, \ldots, a_n$ if $f(X_1, \ldots, X_n)$ is a symmetric or alternating polynomial, respectively.

$s_1, s_2, \ldots, s_n$ defined below are symmetric polynomials of degrees 1, 2, ..., n respectively. They are called the <u>elementary symmetric polynomials</u>.

$$s_1 = X_1 + \cdots + X_n$$
$$s_2 = \textstyle\sum_{i<j} X_i X_j$$
$$\cdots\cdots\cdots\cdots$$
$$s_r = \textstyle\sum_{i_1<i_2<\cdots<i_r} X_{i_1} X_{i_2} \cdots X_{i_r}$$
$$\cdots\cdots\cdots\cdots$$
$$s_n = X_1 X_2 \cdots X_n$$

<u>Elementary symmetric forms</u> of elements $a_1, \ldots, a_n$ of R are $s_1(a_1,\ldots,a_n) = \sum a_i, \ldots, s_n(a_1,\ldots,a_n) = a_1 \cdots a_n$.

THEOREM 2.8.1 Every symmetric polynomial f is expressed as a polynomial in the elementary symmetric polynomials; namely, the set of symmetric polynomials in $R[X_1,\ldots,X_n]$ coincides with $R[s_1, \ldots, s_n]$.

PROOF: For each monomial $m = X_1^{i_1} X_2^{i_2} \cdots X_n^{i_n}$, let d(m) be $(\sum_j i_j, i_1, \ldots, i_n)$. We define two monomials m, m' to be m > m' if d(m) is greater than d(m') in the lexicographical order. We prove the assertion by induction on the maximum d(f) of d(m), m appearing in the symmetric polynomial f of consideration. The least monomial is 1, and therefore if $d(f) \le d(1)$, then $f \in R$ and the assertion is obvious. So we assume that $d(f) > d(1)$. Let $m = X_1^{c_1} \cdots X_n^{c_n}$ be the monomial such that $d(m) = d(f)$. Since, for every $\sigma \in S_n$, σm appears in f (because $\sigma f = f$), we see that $d(m) \ge d(\sigma m)$; hence $c_1 \ge c_2 \ge \cdots \ge c_n$. Set $g = s_n^{c_n} s_{n-1}^{c_{n-1}-c_n} \cdots s_1^{c_1-c_2}$. Then m is the greatest term in g. Hence, letting a be the coefficient of m in f, we see that $d(f-ag) < d(f)$. Therefore we complete the proof, applying the induction hypothesis to f - ag. QED

Next, we observe alternating polynomials. Note first that if the characteristic of R is 2, then $-1 = 1$ and therefore no alternating polynomial exists. The following form D is called the <u>difference product</u> of $X_1, \ldots, X_n$ and is an alternating polynomial unless the characteristic of R is 2.

$$D = \Pi_{i<j} (X_i - X_j)$$

Similarly, the <u>difference product</u> of elements $a_1, \ldots, a_n$ in a ring is $\Pi_{i<j} (a_i - a_j)$. Roughly speaking, above D is the simplest alternating polynomial because of the following fact:

THEOREM 2.8.2 Assume that R is an integral domain of characteristic different from 2.† Then every alternating polynomial f is the product of D and a symmetric polynomial.

In order to prove this, we discuss the symmetric group S_n. An element σ of S_n is called a <u>cyclic</u> permutation of <u>length</u> r if there are mutually distinct elements $i_1, \ldots, i_r$ of $\{1,\ldots,n\}$ such that $\sigma i_j = i_{j+1}$ for $j < r$, $\sigma i_r = i_1$ and $\sigma k = k$ for other k. In this case σ is often denoted by $(i_1,\ldots,i_r)$. $r \geq 2$ unless $\sigma = 1$. In case $r = 2$, we call σ a <u>transposition</u>. For convenience, we understand that every cyclic permutation (r) of length 1 is the identity.

Let us begin with a remark on the multiplication in S_n. When an element σ of S_n is expressed in the form $\begin{pmatrix} a_1 & \cdots & a_n \\ b_1 & \cdots & b_n \end{pmatrix}$, we understand that $\sigma a_i = b_i$ for each i ($\{a_1,\ldots,a_n\}$ and $\{b_1,\ldots,b_n\}$ should be $\{1,\ldots,n\}$). Under the notation, we have the following:

LEMMA 2.8.3 If $\sigma, \tau \in S_n$ and if $\tau = \begin{pmatrix} 1 & 2 & \cdots & n \\ c_1 & c_2 & \cdots & c_n \end{pmatrix}$, then

$$\sigma\tau\sigma^{-1} = \begin{pmatrix} \sigma 1 & \sigma 2 & \cdots & \sigma n \\ \sigma c_1 & \sigma c_2 & \cdots & \sigma c_n \end{pmatrix}$$

PROOF:

$$\tau\sigma^{-1} = \begin{pmatrix} 1 & 2 & \cdots & n \\ c_1 & c_2 & \cdots & c_n \end{pmatrix}\begin{pmatrix} \sigma 1 & \sigma 2 & \cdots & \sigma n \\ 1 & 2 & \cdots & n \end{pmatrix} = \begin{pmatrix} \sigma 1 & \sigma 2 & \cdots & \sigma n \\ c_1 & c_2 & \cdots & c_n \end{pmatrix}$$

QED

LEMMA 2.8.4 Every element σ of S_n is expressed as the pro-

---

† It is sufficient to assume that R is a ring in which 2 is not a zero-divisor. In that case, we have to prove that D is not a zero-divisor in the polynomial ring. Although this is well known, the proof is not easy.

duct of cyclic permutations $\pi_1, \ldots, \pi_\alpha$ such that $\pi_i c \neq c$, $\pi_j c \neq c$ for some $c \in \{1,\ldots,n\}$ implies $i = j$ [i.e., if we express these π_i in the form $(i_1,\ldots,i_r)$ then there is no letter appearing in both of the expressions of distinct π_i and π_j.]

PROOF: Let π_1 be the cyclic permutation $(1,\sigma 1,\ldots,(\sigma^{r_1})1)$ with r_1 such that $\sigma^s 1 \neq 1$ for $s \leq r_1$ and $(\sigma^{r_1+1})1 = 1$. When we constructed $\pi_1, \ldots, \pi_{t-1}$, we define π_t to be $(c,\sigma c,\ldots,\sigma^{r_t}c)$ with c not appearing in any of $\pi_1, \ldots, \pi_{t-1}$ and with r_t defined similarly as r_1 with respect to c. By our construction, we see that if x appears in our expression of $\pi_1, \ldots, \pi_t$, then $\sigma x = (\pi_1 \cdots \pi_t)x$. Now, since n is finite, this procedure terminates, and we obtain the required expression. QED

LEMMA 2.8.5 Every element of S_n is expressed as a product of transposition. Although there are many such expressions, whether the number of factors is odd or even is well determined by the element.

In case the element is the product of an odd number of transpositions, it is called an <u>odd permutation</u>; otherwise, it is called an <u>even permutation</u>.

PROOF: As for cyclic permutations, we have $(a_1,\ldots,a_r) = (a_1,a_r)(a_1,a_{r-1})\cdots(a_1,a_2)$. Therefore we have the first half by Lemma 2.8.4. As for the last assertion, we consider the difference product $D = \Pi_{i<j}(X_i - X_j)$ of $X_1, \ldots, X_n$ over a field K of characteristic zero. Then, for each transposition τ, we have $\tau D = -D$. Therefore σ is an odd permutation if and only if $\sigma D = -D$. QED

In the symmetric group S_n, the set of even permutations forms a normal subgroup of S_n and is called the <u>alternating group</u> of degree n.

PROOF of THEOREM 2.8.2: For each $\sigma \in S_n$, let $s(\sigma)$ be ± 1 such that $\sigma f = s(\sigma)f$. Since $(\sigma\tau)f = \sigma(\tau f) = \sigma(s(\tau)f) = s(\tau)s(\sigma)f$, we have $s(\sigma\tau) = s(\sigma)s(\tau)$ for arbitrary $\sigma,\tau \in S_n$. Set $c = s((1,2))$. Then, for an arbitrary transposition (i,j), $(i,j) = \sigma(1,2)\sigma^{-1}$ with a $\sigma \in S_n$ by Lemma 2.8.3. We have $S((i,j)) = s(\sigma)^2 s((1,2)) = c$. Hence, if τ is an even permutation, then $s(\tau) = 1$ and therefore

$c = -1$. If $i < j$, then as for $h = f(X_1,\ldots,X_{j-1},X_i,X_{j+1},\ldots,X_n)$, we have $h = -h$ because $(i,j)f = -f$. Since 2 is not a zero-divisor in R, we see that $h = 0$. Therefore, by the factor theorem, we see that f is divisible by $X_i - X_j$. Then, adapting our proof of Corollary 2.5.5, we see that f is divisible by D. Let $f = Df^*$. Then for each $\sigma \in S_n$, $\sigma f = s(\sigma)f = s(\sigma)Df^*$, and on the other hand, $\sigma f = (\sigma D)(\sigma f^*) = s(\sigma)D(\sigma f^*)$. Therefore we have $\sigma f^* = f^*$, i.e., f^* is a symmetric polynomial. QED

2.9 INTEGRAL DEPENDENCE

Let S be a subring of a ring R having the same identity. An element a of R is said to be integral over S if there are a natural number n and elements $c_1, \ldots, c_n$ of S such that $a^n + c_1a^{n-1} + \cdots + c_n = 0$. R is said to be integral over S if every element of R is integral over S. In order to observe this notion of integral dependence, we shall make use of the following known fact on linear equations:

LEMMA 2.9.1 Assume that n linear relations $\sum_{j=1}^n c_{ij}a_j = 0$ $(i = 1, \ldots, n)$ holds for elements $a_1, \ldots, a_n$ of a module M over a ring R $(c_{ij} \in R)$. Then, letting D be the determinant $\det(c_{ij})$ of the coefficient matrix, we have $Da_i = 0$ for $i = 1, \ldots, n$.†

PROOF: Using cofactors D_{ij}, we have $\sum_{i=1}^n c_{ij}D_{ik} = 0$ if $j \neq k$, $\sum_{i=1}^n c_{ij}D_{ij} = D$, as is well known. Hence $0 = \sum_i (\sum_j c_{ij}a_j)D_{ik} = Da_k$. QED

THEOREM 2.9.2 Let R and S be as above. Then an element a of R is integral over S if and only if there is a ring T such that $S \subseteq S[a] \subseteq T$ and such that T is finitely generated as an S-module.

PROOF: If a is integral over S, then $a^n + c_1a^{n-1} + \cdots +$

---

† Though many textbooks on determinants deal with those over a field, the theory is generalized to those over a ring. One way is to observe two steps: (i) Every ring is a homomorphic image of an integral domain and (ii) every integral domain is a subring of a field. Another way is to adapt the treatment over a field to the case over a ring.

$c_n = 0$ $(c_i \in S)$ and $S[a] = S + Sa + \cdots + Sa^{n-1}$. Then we may choose T to be $S[a]$. Assume the existence of T and let $b_1, \ldots, b_m$ be such that $T = Sb_1 + \cdots + Sb_m$. Since $a \in T$ and since T is a ring, we have $ab_i = \sum_j c_{ij}b_j$ $(c_{ij} \in S)$. Letting δ_{ij} be Kronecker δ (i.e., $\delta_{ii} = 1$ and $\delta_{ij} = 0$ if $i \neq j$), we have linear equations $\sum_j (a\delta_{ij} - c_{ij})b_j = 0$ $(i = 1, \ldots, m)$. Let D be the determinant $\det(a\delta_{ij} - c_{ij})$. Then $Db_i = 0$ for all j. Since $1 \in T$, we see that $D = 0$. The expansion of D gives us a monic polynomial in a with coefficients in S. Therefore a is integral over S. QED

COROLLARY 2.9.3 Assume that a ring R and its subring S have a common identity. Then the set T of elements a of R, which are integral over S, forms a ring.

This T is called the integral closure of S in R. If $T = S$, then we say that S is integrally closed in R.

PROOF: If $a,b \in T$, then $S[a]$ and $S[a,b]$ are finitely generated S-module, and hence it is integral over S. In particular, ab and $a \pm b$ are integral over S.

COROLLARY 2.9.4 Let S, R', and R^* be rings such that $S \subseteq R' \subseteq R^*$. If R' is integral over S and if R^* is integral over R', then R^* is integral over S. Therefore, under the notation in Corollary 2.9.3, T is integrally closed in R.

PROOF: For each $a \in R^*$, we have $a^n + c_1a^{n-1} + \cdots + c_n = 0$ $(c_i \in R')$. Then the ring $S[c_1,\ldots,c_n,a]$ is finitely generated as an S-module, and hence a is integral over S. QED

Let R be an integral domain with field K of fractions. The integral closure of R in K is called the derived normal ring of R. If R is integrally closed in K, then R is called a normal ring.

THEOREM 2.9.5 Every UFD is a normal ring.

PROOF: Assume that a/b is integral over a UFD R $(a,b \in R)$. We may assume that a and b have no nontrivial common factor. There is a relation $(a/b)^{n-1} + \cdots + c_n = 0$ $(c_i \in R)$. Then $a^n + c_1a^{n-1}b$

$+ \cdots + c_n b^n = 0$; hence $a^n \in bR$. If p is a prime factor of b, then p divides a^n, and hence p divides a. This contradicts our assumtion. Thus b is a unit and $a/b \in R$. QED

EXERCISE 2.3

1. An element e of a ring R is called an <u>idempotent</u> if $e^2 = e \neq 0$.
   (1) Prove that if a ring R is the direct sum of rings $R_1, \ldots, R_n$, then the identity e_i of R_i is an idempotent and the identity of R is the sum of e_i.
   (2) Prove conversely that if the identity of a ring R is the sum of idempotents $e_1, \ldots, e_n$ such that $e_i e_j = 0$ for all i, j such that $i \neq j$, then $R_i = e_i R$ is a ring having e_i as its identity and R is the direct sum of $R_1, \ldots, R_n$.
   (3) Let R, e_i, and R_i be as in (1) above and let M be an R-module. Prove that $e_i M$ is an R_i-module and at the same time an R-module and that M is the direct sum of $e_1 M, \ldots, e_n M$.
2. Does it happen that even if a ring R is the sum of nonzero ideals $I_1, \ldots, I_n$ such that $I_i \cap I_j = 0$ if $i \neq j$, R may not be the direct sum of $I_1, \ldots, I_n$?

EXERCISE 2.4

1. Let p be a prime ideal of a ring R and let S be a subring of R. Prove that $P \cap S$ is a prime ideal of S unless $S \subseteq P$.
2. Let f be a homomorphism of a ring R into a ring S and let P be a prime ideal of S. Prove that $f^{-1}(P)$ is a prime ideal of P unless $fR \subseteq P$.

EXERCISE 2.5

1. In the polynomial ring $R[X]$ over a ring R, a polynomial of positive degree can be a unit. Prove that such can happen if and only if R has a nonzero nilpotent element.

2. In the polynomial ring $R[X]$ over a ring R having no nilpotent elements other than zero, it can happen that $\deg(fg) < \deg f + \deg g$. Show such an example.
3. Consider an ideal I of a ring R and the polynomial ring $R[X_1, \ldots, X_n]$ in n variables over R. Prove that $R[X_1, \ldots, X_n]/IR[X_1, \ldots X_n]$ is isomorphic to the polynomial ring $(R/I)[X_1, \ldots, X_n]$ over R/I.

EXERCISE 2.6

1. Prove that an integral domain D is a UFD if and only if the following two conditions are satisfied:
   (i) Every nonzero element of D is the product of a finite number of irreducible elements.
   (ii) If a $(\in D)$ is a common multiple of two elements b and c and if b and c have no common divisor except units, then a is a multiple of bc.
2. Let E, v, W be as in the definition of a Euclid ring. We define a mapping v' of E into W by:

   $$v'a = \inf \{vb \mid b \text{ is an associate of } a\}$$

   Prove that E, v', W satisfy the conditions (i), (ii) in the definition, and also the following:

   (iii) If b is a multiple of a and if $b \neq 0$, then $v'b \geq v'a$.

   We note here that often in the literature the notion of a Euclid ring is defined assuming (iii), too.

EXERCISE 2.7

1. Show that every module is naturally a module over the ring Z of rational integers.
2. Prove that the field of rational numbers is not a free module over Z.
3. Prove that if M is an Artin module over a ring R and if N is a submodule, then

   $$\text{length } M = \text{length } N + \text{length } (M/N)$$

4. Let M and N be modules over a ring R. Show by an example that a submodule L of the direct sum $M \oplus N$ need not be $(L \cap M) \oplus (L \cap N)$. Prove, on the other hand, that if R and R' are rings, then an ideal A of the direct sum $R \oplus R'$ coincides with $(A \cap R) \oplus (A \cap R')$.
5. Let R and R' be Artin rings. Prove that the direct sum $R \oplus R'$ is an Artin ring.

EXERCISE 2.8

1. Let K be a field of characteristic different from 2. Consider the polynomial ring $K[X_1,\ldots,X_n]$ and the alternating group A_n on the set $\{1,\ldots,n\}$. Prove that, for an element $f \in K[X_1,\ldots, X_n]$, $\sigma f = f$ for every $\sigma \in A_n$ if and only if f is the sum of a symmetric form and an alternating form (or 0). [<u>Hint</u>: Consider $f - \tau f$ with transposition $\tau = (1,2)$.]
2. Confirm that $\sum_{i=1}^{n} (X - a_i) = X^n + c_1X^{n-1} + \cdots + c_n$ if $-c_1$, c_2, ..., $(-1)^r c_r$, ..., $(-1)^n c_n$ are the elementary symmetric forms of degrees 1, 2, ... r, ..., n, respectively, of a_1, ..., a_n.
3. Prove that if a polynomial f is symmetric, then f is a sum of homogeneous symmetric forms.

EXERCISE 2.9

1. (<u>Irreducibility theorem of Eisenstein</u>) Let R be an integral domain, K its field of fractions, and $f(X) = c_0X^n + c_1X^{n-1} + \cdots + c_n$ $(c_i \in R)$ a polynomial such that its coefficients c_0, ..., c_n have no nontrivial common factor. Prove that $f(X)$ is irreducible in $R[X]$ if there is a prime ideal P such that (i) $c_0 \notin P$, (ii) $c_1, \ldots, c_n \in P$, and (iii) $c_n \notin P^2$.

CHAPTER 3

ALGEBRAIC EXTENSIONS OF FINITE DEGREES

In this chapter, we study a theory of algebraic extension fields of finite degrees, including the so-called fundamental theorem of Galois. Furthermore, we give some applications of the theory, including the existence and the uniqueness of the algebraic closure of a given field.

3.1 ALGEBRAIC EXTENSIONS

When K is subfield of a field L and S is a subset of L, the smallest field containing K and S exists (it is the intersection of all subfields of L containing K and S). The subfield is called the field <u>generated</u> by S over K and is denoted by $K(S)$. If $S = \{a_1, \ldots, a_n\}$, then $K(S)$ is denoted also by $K(a_1, \ldots, a_n)$. If K and M are subfields of L, then $K(M)$ $(= M(K))$ may be denoted by $K \vee M$. On the other hand, a field L' containing a field K is called an <u>extension field</u> of K or simply an <u>extension</u> of K.

When K is an integral domain, the subring P of K generated by the identity of K is the smallest subring of K. This P is called the <u>prime integral domain</u> of K. If, furthermore, K contains the field of fractions F of P (so does K if K is a field), then F is called the <u>prime field</u> of K. F is the smallest subfield of K. Note that:

PROPOSITION 3.1.1 The characteristic p of the integral domain K is either zero or a prime number. Furthermore:

(1) If $p = 0$, then P is isomorphic to the ring Z of rational integers and F is isomorphic to the field Q of rational numbers.

Hence, hereafter, we identify P and F with Z and Q, respectively, in the characteristic zero case.

(2) If p is a prime number, then P is a field consisting of p elements and is isomorphic to Z/pZ. Hence P = F, and we identify P = F with Z/pZ is this case.

PROOF: Assume that p = ab with natural numbers a, b. Then (a-times identity)(b-times identity) = 0. Since K is an integral domain, one of the factors must be zero. Then, by the definition of the characteristic, we see that p = a or b, which proves that p is either a prime number or zero. Observe the mapping f of Z into K such that $fn = n \cdot 1$. Then f is a homomorphism and fZ = P. Therefore we see that P = Z/pZ. Hence (1) is immediate. If $p \neq 0$, then, since Z is a Euclid ring, we see that pZ is a maximal ideal of Z (Theorem 2.6.3). This implies that Z/pZ is a field. QED

Assume that S is a subring of an integral domain K. (1) An element a of K is said to be <u>algebraic</u> over S if there are a natural number n and elements $c_0, \ldots, c_n$ of S such that $c_0a^n + c_1a^{n-1} + \cdots + c_n = 0$, $c_0 \neq 0$. (2) If every element of K is algebraic over S, then we say that K is <u>algebraic</u> over S. (3) For an element of K or K itself, we say that it is <u>transcendental</u> over S unless it is algebraic over S.

Note first that in (1) above, if c_0 is a unit in S, then a is integral over S. Therefore if S is a field, "algebraic" and "integral" mean the same. In such a case, we normally use the term "algebraic." Note also that, in (1) above, c_0a is integral over S as is easily seen, and conversely, if ca $(a \in K,\ 0 \neq c \in S)$ is integral over S, then a is algebraic over S. By this fact, we can adapt results in Sec. 2.9 to the algebraic case. For instance, the set S* of algebraic element of K over S forms a subring of K. This S* is called the <u>algebraic closure</u> of S in K. If S* = S, then we say that S is <u>algebraically closed</u> in K.

Let L be an extension field of a field K. Then $\text{length}_K L$ is called the <u>degree</u> of the extension, and we denote it by [L : K]. Note that if a $(\in L)$ is transcendental over K, then $[L : K] \geq [K(a)$

: K] = ∞. Therefore, if [L : K] is finite, then L must be algebraic over K; hence L is called an algebraic extension of finite degree, or simply, a finite algebraic extension, of K.

A field K is called a finite field if #(K) is finite; otherwise K is called an infinite field.

THEOREM 3.1.2 Let P and p be the prime field and the characteristic, respectively, of a field K. If K is a finite field, then setting $e = [K : P]$ we have $\#(K) = p^e$. Furthermore, every element of K is a root of the polynomial $X^{p^e} - X$.†

PROOF: Since #(K) is finite, p is a prime number and #(P) = p by Proposition 3.1.1. Let $b_1, \ldots, b_e$ be a linearly independent base for K over P. Then every element of K is expressed uniquely as a linear combination $\sum c_i b_i$ of the b_i with coefficients in P. Therefore we see that $\#(K) = p^e$. Let K* be the set of nonzero elements of K. Then K* is a group consiting of $p^e - 1$ elements. Hence every element of K* is a root of $X^m - 1$ with $m = p^e - 1$, which proves the last assertion. QED

A remarkable fact on a ring of prime characteristic p is that $(a + b)^p = a^p + b^p$, which is easily seen since binomial coefficients $\binom{p}{1}, \ldots, \binom{p}{r}, \ldots, \binom{p}{p-1}$ are multiples of p. This fact can be generalized easily to the following form:

LEMMA 3.1.3 Assume that the characteristic p of a ring R is a prime number. If $a_1, \ldots, a_n \in R$ and if q is a power p^s of p, then

$$(a_1 + \cdots + a_n)^q = a_1{}^q + \cdots + a_n{}^q$$

The following result is basic in computing degrees of extensions.

THEOREM 3.1.4 Assume that L, K are algebraic extensions of fields K, M of degrees m, n, respectively. Then the degree of extension of L over M is equal to mn.

---

† By virtue of the remark at the end of Sec. 2.6, this polynomial has at most p^e roots counting multiplicies. On the other hand, $\#(K) = p^e$. Therefore all roots of $X^{p^e} - X$ are simple roots and $X^{p^e} - X = \prod_{a \in K} (X - a)$.

PROOF: Let $b_1, \ldots, b_m$ and $c_1, \ldots, c_n$ be linearly independent bases for L over K and K over M, respectively. It suffices to show that $\{b_i c_j \mid i = 1,\ldots,m;\ j=1, \ldots, n\}$ form a linearly independent base for L over M. If a L, then $a = \sum_i e_i b_i$ with $e_i \in K$. Since each e_i is a linear combination of the c_j with coefficients in M, we see that $L = \sum_{i,j} b_i c_j M$. Assume that $\sum_{i,j} f_{ij} b_i c_j = 0$ with $f_{ij} \in M$. Since the b_i are linearly independent over L, we see that $\sum_j f_{ij} c_j = 0$ for every fixed i. Since the c_j are linearly independent over M, we see that every f_{ij} is zero. QED

COROLLARY 3.1.5 A field L containing a field K is a finite algebraic extension of K if and only if there are algebraic elements $a_1, \ldots, a_n$ over K such that $L = K(a_1,\ldots,a_n)$.

The following result is important, not only for algebraic extensions of fields but also for integral dependence of rings:

THEOREM 3.1.6 Assume that K is an integral domain and that S is a subring of K such that K is integral over S.

(1) If one of K and S is a field, then the other is also a field.

(2) If an element b of S has its inverse b^{-1} in K, then $b^{-1} \in S$.

PROOF: Assume first that S is a field. If $0 \neq a \in K$, then $a^n + c_1 a^{n-1} + \cdots + c_n = 0$ with $c_i \in S$. Since K is an integral domain, we may assume that $c_n = 0$. Since S is a field, $c_n^{-1} \in S$; hence $a^{-1} = c_n^{-1}(a^{n-1} + c_1 a^{n-1} + \cdots + c_{n-1})$. Thus $a^{-1} \in K$ and K is a field. Assume now that K is a field. If $0 \neq b \in S$, then $b^{-1} \in K$; hence b^{-1} is integral over S. Therefore $(b^{-1})^m + e_1(b^{-1})^{m-1} + \cdots + e_m = 0$ with $e_j \in S$. Then $b^{-1} = -(e_1 + e_2 b + \cdots + e_m b^{m-1}) \in S$, which proves that S is a field. (2) is proved similarly. QED

Now we go back to finite algebraic extensions.

THEOREM 3.1.7 Assume that a field L is generated by a single element a over a field K and that a is algebraic over K.

Consider the ideal $P = \{g(X) \in K[X] \mid g(a) = 0\}$ in the polynomial ring $K[X]$. Then, (1) P is generated by a monic irreducible polynomial, say $f(X) = X^n + c_1X^{n-1} + \cdots + c_n$ ($c_i \in K$); (2) its degree n coincides with $[L : K]$; (3) $L = K[a]$; and (4) $1, a, \ldots, a^{n-1}$ form a linearly independent base for L over K.

The polynomial $f(X)$ is determined uniquely by the pair (K,a), and is called the minimal polynomial for a over K.

PROOF: $K[a]$ is a field by Theorem 3.1.6, and hence $L = K[a]$. Obviously, $K[a] \cong K[x]/P$; hence P is a prime ideal. Since $K[X]$ is a principal ideal domain by Theorem 2.6.3, we see the existence of $f(X)$. Then, since $a^n = -(c_n + c_{n-1}a + \cdots + c_1a^{n-1}) \in \sum_{i=0}^{n-1} a^iK$, we see easily that $L = K[a] = \sum_0^{n-1} a_iK$. If $1, a, \ldots, a^{n-1}$ are linearly dependent over K, then P must contain elements of degree smaller than n, which is impossible. QED

THEOREM 3.1.8 If a field $L = K(a_1,\ldots,a_n)$ is algebraic over a field K, then L is isomorphic to $K[X_1,\ldots,X_n]/M$ with maximal ideal M of the polynomial ring $K[X_1,\ldots,X_n]$, which is generated by n elements $f_1, \ldots, f_n$ such that (i) f_i is in $K[X_1,\ldots,X_i]$ and is monic as a polynomial in X_i and (ii) $f_i(a_1,\ldots,a_{i-1},X_i)$ is the minimal polynomial for a_i over $L_i = K(a_1,\ldots,a_{i-1})$ for each i.

PROOF: Let M be the kernel of the K-homomorphism h of $K[X_1,\ldots,X_n]$ into L such that $hX_i = a_i$ for all i. Then $K[a_1,\ldots,a_n] \cong K[X_1,\ldots,X_n]/M$. By Theorem 3.1.6, $K[a_1,\ldots,a_n]$ is a field, and $L = K[a_1,\ldots,a_n] \cong K[X_1,\ldots,X_n]/M$. We see immediately that these f_i are in M. On the other hand, letting M_i be the ideal of $K[X_1,\ldots,X_i]$ generated by $f_1, \ldots, f_i$, we see that

$$K[X_1,\ldots,X_i]/M_i \cong L_{i-1}[X_i]/(f_i(a_1,\ldots,a_{i-1},X_i)L_{i-1}[X_i]) \cong L_i$$

by induction on i. Theorefore the ideal generated by $f_1,\ldots,f_n$ is maximal, and hence it coincides with M. QED

From this result, we derive a lemma which we need in the next section.

LEMMA 3.1.9 Assume that L is a finite algebraic extension of a field K. If σ is an isomorphism of K onto a field K', then there is a field L' such that σ is extended to an isomorphism σ' of L onto L' (i.e., $K' \subseteq L'$ and $\sigma'|_K = \sigma$).

PROOF: $L = K[X_1,\ldots,X_n]/M$ with a maximal ideal M of the polynomial ring $K[X_1,\ldots,X_n]$ as was shown above. σ is extended to an isomorphism σ^* of $K[X_1,\ldots,X_n]$ onto $K'[X_1,\ldots,X_n]$ so that $\sigma^*X_i = X_i$. Then σ^*M is a maximal ideal, and therefore $L' = K'[X_1, \ldots X_n]/\sigma^*M$ and the naturally induced isomorphism σ' of L onto L' are the required QED

3.2 SPLITTING FIELDS

For a polynomial $f(X)$ in a variable X over a field K, a field L containing K is called a splitting field for $f(X)$ if $f(X)$ is the product of some polynomials of degree one over L. If L^* is minial among such splitting fields, then we call L^* a minimal splitting field. Note that $f(X) = a(X-c_1)\cdots(X-c_n)$ in a splitting field L $(c_i \in L)$ and that L is minimal if and only if $L = K(c_1,\ldots,c_n)$.

THEOREM 3.2.1 If $f(X)$ is a polynomial over a field K, then there is a splitting field for $f(X)$.

PROOF: We prove this by an induction on the degree d of $f(X)$. If $d = 0$ or 1, then K is a splitting field for $f(X)$. Assume that $d > 1$, and let $g(X)$ be an irreducible factor of $f(X)$ in $K[X]$. Then the field $L' = K[X]/g(X)K[X]$ is a finite algebraic extension of K in which the class a of X is a root of $g(X)$. Then $f(X) = (X - a)h(X)$ as a polynomial over L', and a splitting field L for $h(X)$ over L' is a splitting field for $f(X)$ over K, and L exists by our induction hypothesis. QED

Now we come to the uniquness theorem of minimal splitting fields. Let K and $f(X)$ be as above [deg $f(X) \geq 1$, for simplicity], and let $g(X)$ be an irreducible factor of $f(X)$ in $K[X]$. Then:

Theorem 3.2.2 If L and L^* are minimal splitting fields for $f(X)$ over K, and if b, b^* are roots of $g(X)$ in L, L^* respectively, then there is a K-isomorphism ψ^* of L onto L^* such that $\psi^*b = b^*$. More generally, if σ is an isomorphism of K onto a field

K' and if L, L' are minimal splitting fields for $f(X)$, $\sigma f(X)$ (σ being extended so that $\sigma X = X$), then for given roots b, b' of $g(X)$, $\sigma g(X)$, respectively, there is an isomorphism σ' of L onto L^* such that σ' is an extension of σ and $\sigma' b = b'$. Conversely, if σ'' is an isomorphism of L onto L' which extends σ, then $\sigma'' b$ is a root of $\sigma g(X)$.

PROOF: It suffices to prove the last half, and we use an induction on $[L : K]$. Since $K[b] \cong K[X]/g(X)K[X]$ and $K'[X]/\sigma g(X)K'[X] \cong K'[b']$, we see that σ is extended to an isomorphism σ^* of $K[b]$ onto $K'[b']$; hence by our induction assumption, we see that σ^* extends to an isomorphism of L onto L'. This proves the assertion, except for the last comment, which is obvious because σ'' induces an isomorphism of $K[b]$ onto $K[\sigma'' b]$. QED

Let K be a subfield of a field L. Elements a, b of L are said to be conjugate to each other over K, if a and b are roots of a common irreducible polynomial over K. Note that the proof above shows the following proposition.

PROPOSITION 3.2.3 a, b ($\in L$) are conjugate to each other over K if and only if there is a K-isomorphism of a field containing L to itself which maps a to b.

In view of this, we define that two subsets M and M'' of an algebraic extension L of a field K are conjugate to each other if there is a K-isomorphism τ of a field containing L onto itself such that $\tau M = M'$.

At the moment, we are constructing a theory of finite algebraic extensions without using the existence of algebraic closures (which will be proved in Sec. 3.11). For our purpose, we consider sufficiently large fields in the following sense: If $L = K(a_1, \ldots, a_n)$ is a finite algebraic extension of a field K, then, letting $f_i(X)$ be the minimal polynomial for a_i over K, L is a subfield of minimal splitting field S for $f_1(X) \cdots f_n(X)$. In order to observe algbraic properties of L, we may choose any S because of Theorem 3.2.2. If $L' = K(b_1, \ldots, b_m)$ is another finite algebraic extension of K, then taking a similar polynomial $g(X)$ as $f(X)$, both L and L' may be identified with subfields of a minimal splitting field for $f(X)g(X)$ over K. Thus, when we are observing a finite number of finite algebraic extensions of a given field K, we may embed them via isomorphisms into a minimal splitting field S^* of a certain polynomial over K. Such an S^* is called a sufficiently large field (with respect to these finite algebraic extensions). Thus, throughout Secs. 3.3 to 3.11, when we deal with a finite number of finite algebraic extensions of a field K, we assume that they are subfields of a sufficiently large field.

3.3 SEPARABILITY

Let a be an element of an algebraic extension field L of a field K. If the minimal polynomial $f_a(X)$ for a over K has a as a simple root, then we say that a is separable over K; otherwise it is inseparable. In connection with this definition of separability, we define that a polynomial $f(X)$ over a field is said to be separable if $f(X)$ has no multiple root. On the other hand, if $f_a(X)$ has only one root (allowing multiplicity), then we say that a is purely inseparable over K. (a is separable and purely inseparable over K only when a is in K.) Note that a is separable over K if there is a polynomial $f(X)$ over K such that a is a simple root of $f(X)$, because $f_a(X)$ is a factor of such an $f(X)$. If every element of L is separable over K, then we say that L is separable, or separably algebraic, over K; otherwise we say that L is inseparable over K. If every element of L is purely inseparable over K, then we say that L is purely inseparble over K. K is said to be perfect if every algebraic extension of K is separable over K.

We define here derivatives of polynomials. For a polynomial $f(X) = a_n X^n + \cdots + a_0$ over a ring K ($a_i \in K$), its derivative is defined to be $na_n X^{n-1} + \cdots + ra_r X^{r-1} + \cdots + a_1$ and is denoted by $\frac{d}{dX} f(X)$. Then:

LEMMA 3.3.1 Let $f(X)$ be a polynomial over a field K, and let a be an element of an extension field L of K. Then a is a multiple root of $f(X)$ if and only if a is a common root of $f(X)$ and its derivative $\frac{d}{dX} f(X)$.

Before proving this, we note that the operator $\frac{d}{dX}$ enjoy the properties:

$$\frac{d}{dX}(f(X) + g(X)) = \frac{d}{dX} f(X) + \frac{d}{dX} g(X)$$

$$\frac{d}{dX} f(X)g(X) = f(X)\frac{d}{dX} g(X) + g(X)\frac{d}{dX} f(X)$$

for arbitrary polynomials $f(X)$, $g(X)$ in a variable X over K. (The proof is easy and we omit it.) Therefore we have

$$\frac{d}{dX}(f_1(X)\cdots f_m(X)) = \sum_{i=1}^{m} f_1(X)\cdots f_{i-1}(X)f_{i+1}(X)\cdots f_m(X)\frac{d}{dX} f_i(X)$$

as is easily seen by induction on m.

PROOF of the LEMMA: Note first that derivatives do not depend on the choice of the field containing the coefficients. Let $f(X) = (X - a)^e g(X)$ with $g(a) \neq 0$. If a is a multiple root, then $e \geq 2$ and $\frac{d}{dX} f(X) = e(X - a)^{e-1}g(X) + (X - a)^e \frac{d}{dX} g(X)$. Therefore a is a common root of $f(X)$ and its derivative. Conversely, assume that a is a common root of $f(X)$ and its derivative. Then $e \geq 1$. If $e = 1$, then by the computation above, we see that a is not a root of $\frac{d}{dX} f(X)$ and we have $e \geq 2$. QED

THEOREM 3.3.2 Every field K of characteristic zero is perfect.

PROOF: Let $f(X)$ be an irreducible polynomial in X over K. If $f(X)$ has a multiple root, then $f(X)$ and $\frac{d}{dX} f(X)$ must have a common factor by Proposition 2.6.7. Since $f(X)$ is irreducible and since the degree of the derivative is less than that of $f(X)$, we see that $\frac{d}{dX} f(X)$ must be zero. This happens only if $\deg f(X) = 0$. QED

The argument above can be applied to the case where the field K is of characteristic $p \neq 0$. Namely, in that case, $\frac{d}{dX} f(X) = 0$ if and only if $f(X)$ is a polynomial in X^p. $f(X) = a_0 + a_1X^p + \cdots + a_nX^{np}$ is $(b_0 + b_1X + \cdots + b_nX^n)^p$ if each b_i is the p-th <u>root</u> of a_i (i.e., $a_i = b_i^p$) by Lemma 3.1.3. Making use of this, we have:

THEOREM 3.3.3 An irreducible polynomial $f(X)$ in X over a field K of characteristic $p \neq 0$ has a multiple root if and only if $f(X)$ is a polynomial in X^p. Let $q = p^e$ be the power of p such that $f(X)$ is a polynomial in X^q, but not in X^{pq}. Then (i) the q-th power c^q of each root c of $f(X)$ is separable over

K, (ii) each root of $f(X)$ is separable over the extension field of K generated by the q-th roots of the coefficients of $f(X)$ over K, and (iii) the multiplicity of every root (in a splitting field) is q.

PROOF: The first half was proved already. As for (i), let $g(X)$ be the polynomial such that $g(X^q) = f(X)$. Then $g(X)$ is irreducible and not a polynomial in X^p. Therefore $g(X)$ has no multiple root. In order to prove (ii) and (iii), let us observe $b_0 + b_1X + \cdots + b_nX^n$ over the field $L = K(b_0, \ldots, b_n)$ (under the symbols given above). If this were reducible, let $c_0 + c_1X + \cdots + c_rX^r$ $(1 \le r < n)$ be a factor. Since $b_i^p \in K$, we see that $c_j^p \in K$ by virtue of Lemma 3.1.3, and we see that $c_0^p + c_1^pX^p + \cdots + c_r^pX^{rp}$ is a factor of $f(X)$ over K, which contradicts the irreducibility of $f(X)$. Thus we have (ii) and (iii) easily by induction on e. QED

COROLLARY 3.3.4 If $f(X)$ is an irreducible polynomial in X over a field K, then the multiplicities of roots of $f(X)$ are equal to each other. In particular, if $f(X)$ has a simple root, then $f(X)$ is separable. On the other hand, if a root of $f(X)$ is purely inseparable over K and if K is of characteristic $p \neq 0$, then the degree of $f(X)$ is a power of p.

THEOREM 3.3.5 A field K of characteristic $p \neq 0$ is perfect if and only if every element a of K has its p-th root in K.

PROOF: If a $(\in K)$ has no p-th root in K, then $X^p - a$ is irreducible over K. Indeed, in a splitting field L for $X^p - a$, we have $X^p - a = (X - b)^p$. Therefore, if $X^p - a$ is reducible over K, then for some e $(1 \le e < p)$, $(X - b)^e$ must be a polynomial over K. Then observing its coefficient of X^{e-1}, we see that $eb \in K$. This contradicts the assumption. Conversely, if every element of K has its p-th root in K, then the argument before Theorem 3.3.3 is applied and K is perfect. QED

REMARK: As is seen in the proof above, in a field K of characteristic $p \neq 0$, if an element a has its p-th root, then the p-th root is unique. Hence the uniqueness is true for p^e-th roots $(e = 1, 2, \ldots)$.

THEOREM 3.3.6 Every finite field K is a perfect field.

PROOF: Let L be a finite algebraic extension of K. Then L is a finite field. Set $q = \#(L)$. Then, by Theorem 3.1.2, every element of L is a root of $X^q - X$, and it is a simple root by the footnote to the theorem. Therefore L is separable over K. QED

In connection with this last result, we add here the following:

THEOREM 3.3.7 If K is a finite field, then the multiplicative group $K^* = K - \{0\}$ is a cyclic group.

PROOF: For every natural number n, the equation $X^n = 1$ has at most n solutations in K^*; hence the assertion follows from Lemma 1.6.1. QED

Another remark we should add here is:

PROPOSITION 3.3.8 Let L be a finite algebraic extension field of a field K of characteristic $p \neq 0$.

(1) If L is inseparable over K, then p divides $[L : K]$.

(2) If L is purely inseparable over K, then $[L : K]$ is a power of p.

PROOF: This follows from Theorem 3.1.4, Theorem 3.3.3, and Corollary 3.3.4.

3.4 SIMPLE EXTENSIONS

A field L is said to be a <u>simple extension</u> of a field K if L is generated by a single element over K.

THEOREM 3.4.1 Let $a_1, \ldots, a_n$ be elements of an algebraic extension field of a field K. If $a_1, \ldots, a_{n-1}$ are separable over K, then $L = K(a_1,\ldots,a_{n-1},a_n)$ is a simple extension of K.

PROOF: If K is a finite field, then by Theorem 3.3.7, the multiplicative group of L is generated by an element a, and $L = K(a)$. Assume now that K is an infinite field. It suffices to prove the case where $n = 2$. Let $f_i(X)$ be the minimal polynomial for a_i over K for each $i = 1, 2$, and let $a_i = a_{i1}, a_{i2}, \ldots, a_{is_i}$ be the roots of $f_i(X)$ in a splitting field for $f_1(X)f_2(X)$. Let c be an element of K which is not equal to any of $(a_{2i} - a_{2j})/(a_{1k} - a_{1m})$

$(k \neq m)$ and set $b = a_2 + ca_1$. Then $a_2 = b - ca_1$, and $f_2(b - ca_1) = 0$. Therefore a_1 is a common root of $f_1(X)$ and $g(X) = f_2(b - cX)$. Since $f_1(X) = \Pi_{i=1}^{s_1}(X - a_{1i})$ and $g(X) = \Pi_{g=1}^{s_2}(b - cX - a_{2j}) = (-c)^{s_2} \cdot \Pi_j (X - a_1 - c^{-1}(a_2 - a_{2j}))$, we see that, by our choice of c, $X - a_1$ is the largest common factor of $f_1(X)$ and $g(X)$. But the largest common factor is computed by the Euclid algorithm. Hence the fact that the coefficients of $f_1(X)$ and $g(X)$ are in $K(b)$ implies that a_1 is in $K(b)$. Then $a_2 = b - ca_1 \in K(b)$. Thus $K(a_1,a_2) = K(b)$. QED

THEOREM 3.4.2 For a field L which is a finite algebraic extension of a field K, the following four conditions are equivalent to each other:

(1) L is separable over K.

(2) L is generated by separable elements $a_1, \ldots, a_n$ over K.

(3) $L = K(a)$ with a separable element a.

(4) $\#(I) = [L : K]$, where I is the set of K-injections (i.e., injective K-homomorphisms) of the field L into a sufficiently large algebraic extension Ω of K.

Note here that, as is shown in the proof below, if L is inseparable over K, then $\#(I) < \text{length}_K L$.

PROOF: Assume first that $L = K(a)$ with an element a. Let $a = a_1, \ldots, a_d$ $(d = [L : K])$ be the roots of the minimal polynomial $f(X)$ for a over K. If $\sigma \in I$, then a^σ must be a root of $f(X)$. Conversely, for each a_i, there is σ $(\in I)$ such that $a^\sigma = a_i$ in view of the natural isomorphism $K(a) \cong K[X]/f(X)K[X] \cong K(a_i)$. Therefore $\#(I) =$ [the number of distinct roots of $f(X)$], which proves the equivalence of (3) and (4) under the assumption that $L = K(a)$. Now, assume that (2) holds. Then $L = K(a)$ for an a by Theorem 3.4.1. In order to show (3), we may assume that $n = 2$. Let I_1 be the set of K-injections of $K(a_1)$ into Ω. Then $\#(I_1) = [K(a_1) : K]$. Since a_2 is separable over $K(a_1)$, we see that, for each $\sigma \in I_1$, the number of element of I which are extensions of σ coincides with $[L : K(a_1)]$ [applying the same as above to the extension L over $K(a_1)$]. Therefore we see that $\#(I) = \#(I_1) \times [L : K(a_1)] =$

$[L : K]$ (see Sec. 3.1). Therefore a is separable over K. Thus we see the implications (1) $\Rightarrow$ (2) $\Rightarrow$ (3) $\Rightarrow$ (4). Assume now that (4) holds good. Assume for a moment that an element b of L is inseparable over K. Let a_i be such that $L = K(b,a_1,\ldots,a_m)$, and let I_j be the set of K-injections of $L_j = K(b,a_1,\ldots,a_j)$ into Ω, for each $j = 0, 1, \ldots, m$. As was shown above, $\#(I_0) < [L_0 : K]$. Furthermore, for each $j \geq 1$, we see that the number of elements of I_j which extends a fixed element of I_{j-1} is at most $[L_j : L_{j-1}]$. Therefore we see that $\#(I_j) \leq \#(I_0) \times [L_1 : L_0]\cdots[L_j : L_{j-1}] < [L_j : K]$. In particular, $\#(I) < [L : K]$, which contradicts (4). Thus (4) implies (1). QED

COROLLARY 3.4.3 If a field L is algebraic over a field K, then the set L_s of separable elements of L over K forms a field. L is purely inseparable over L_s.

This L_s is called the <u>separable closure</u> of K in L. $[L_s : K]$ and $[L : L_s]$ are called the <u>degrees of separability</u> and <u>inseparability</u> of L over K and are denoted by $[L : K]_s$ and $[L : K]_i$, respectively.

COROLLARY 3.4.4 Let L be a finite algebraic extension field of a field K, and let I be the set of K-injections of the field L into a sufficiently large algebraic extension Ω. Then $\#(I)$ coincides with the degree of separability of L over K.

PROOF: Each element of I is uniquely determined by its restriction on the separable closure L_s. QED

THEOREM 3.4.5 A finite algebraic extension field L of a field K is a simple extension if and only if there are only a finite number of intermediate fields between K and L.

PROOF: Assume first that $L = K(a)$ with $a \in L$. Let $f(X)$ be the minimal polynomial for a over K. If M is an intermediate field, then $L = M(a)$ and the minimal polynomial $f_M(X)$ for a over M must be a factor of $f(X)$. Let M' be the field generated by the coefficients of $f_M(X)$ over K. Then $M' \subseteq M$, and $f_M(X)$ is irreducible over M'. Since $L = M'(a)$, we see that $[L : M'] = \deg f_M(X) = [L : M]$. This implies that $M = M'$. Thus intermediate fields correspond in one-one way to some of monic factors of $f(X)$, which exist only in a finite number. Conversely, assume that L is not a simple extension of K. Let L_s be the separable closure of K in L. Then Theorem 3.4.1 implies that L is not a simple extension of L_s. Let $a_1, \ldots, a_r$ be such that $L = L_s(a_1,\ldots,a_r)$ and such that L cannot be generated by any $(r - 1)$ elements over L_s $(r \geq 2)$. Let p be the characteristic of K ($p \neq 0$ by Theorem 3.3.2), and set $K^* = L_s(a_1^p,\ldots,a_r^p)$. Then $[L : K^*] = p^r$. [Indeed, setting $L_j = L_s(a_1,\ldots,a_j,a_{j+1}^p,\ldots,a_r^p)$; if $L_j = L_{j+1}$ for some j, we have $a_{j+1} \in L_j$ and $a_{j+1}^p \in L_s(a_1^p,\ldots,a_j^p,a_{j+1}^{p^2},\ldots,a_r^{p^2})$. Hence $a_{j+1} \in L_j = L_s(a_1,\ldots,a_j,a_{j+1}^p,\ldots a_r^p) \subseteq L_s(a_1,\ldots,a_j,a_{j+1}^{p^2},a_{j+2},\ldots, a_r^p)$. Repeating the same, we see that $a_{j+1} \in L_s(a_1,\ldots,a_j,a_{j+1}^{p^n},a_{j+2}^p, \ldots,a_r^p)$ for every n. But, for a sufficiently large n, $a_{j+1}^{p^n} \in L_s$ (Theorem 3.3.3) and we have a contradiction to the minimality of r.] Since L_s is not perfect, L_s is not a finite field (Theorem 3.3.6). For each $c \in L_s$, set $y_c = a_1 + ca_2$ and $M_c = K(y_c)$. Since $y_c^p \in K^*$, $[M_c : K^*] = p$. Since $K^*(y_c,y_{c'}) = K^*(a_1,a_2)$ for every pair of c, c' $(c \neq c')$, we see that all M_c are different from each other. QED

3.5 NORMAL EXTENSIONS

An algebraic extension field L of a field K is called a <u>normal extension</u> of K if it enjoys the property that if $a \in L$ and if $f(X)$ is the minimal polynomial for a over K, L is a splitting field for $f(X)$. Namely, L is normal over K if and only if for each $a \in L$, all conjugates of a over K (in a sufficiently large algebraic extension of L) are in L. A separable, normal extension is called a <u>Galois extension</u> of K.

THEOREM 3.5.1 A finite algebraic extension field $L = K(a_1,\ldots, a_n)$ of a field K is a normal extension of K if and only if every conjugate of each a_i is in L.

PROOF: The only if part is obvious and we observe the if part.

Let $f_i(X)$ be the minimal polynomial for a_i over K. Then the assumption implies that L is a minimal splitting field for $f(X) = f_1(X)\cdots f_n(X)$. Let $b \in L$ and let Ω be a sufficiently large algebraic extension of L. If b' is a conjugate of b over K, then there is a K-injection σ of K(b) into Ω such that $b^\sigma = b'$. If we extend σ to a K-injection of L into Ω (Lemma 3.1.9), then L^σ must be a minimal splitting field for f(X). Therefore $L^\sigma = L$ and $b' \in L$. QED

In general, if L is a ring containing a ring K, then the set $\mathrm{Aut}_k L$ of K-automorphisms of L forms a group. Our main interest in this text is in the case where L is a normal extension of a field K. In such a case, $\mathrm{Aut}_K L$ is called the Galois group of the extension L/K and is denoted by G(L/K). If f(X) is a polynomial in X over a field K, then letting L be a minimal splitting field for f(X) over K, the group $\mathrm{Aut}_K L = G(L/K)$ is called the Galois group of f(X) over K.

If L is a finite, normal extension of a field K, then $L^\sigma = L$ for every K-injection σ of L into a sufficiently large algebraic extension Ω of L, as was seen above. Furthermore, as we saw in Corollary 3.4.4, the number of such injections coincides with $[L : K]_s$. Therefore:

THEOREM 3.5.2 Let L be a finite normal extension of a field K and let L_s be the separable closure of K in L. Then

(1) $\#(\mathrm{Aut}_K L) = [L_s : K] = [L : K]_s$.

(2) L_s is a Galois extension of K and G(L/K) is naturally isomorphic to $G(L_s/K)$ [each element of G(L/K) corresponds to its restriction on L_s].

Furthermore, in view of our proof of Theorem 3.5.1, we have:

THEOREM 3.5.3 If a is an element of a finite normal extension L of a field K, then the set of conjugates of a (in a sufficiently large algebraic extension of K) is $\{a^\sigma \mid \sigma \in G(L/K)\}$. This is true also for subsets of L.

3.6 INVARIANTS OF A FINITE GROUP

We say that a group G acts on a ring R if there is given a homomorphism ψ of G into the automorphism group Aut R so that each element of G induces an automorphism of R by defining $a^\sigma = a^{\psi\sigma}$. In this case, an element a of R is called an invariant of G or a G-invariant if $a^\sigma = a$ for every σ in G. The set of G-invariants in R forms a subring of R, which is called the ring of G-invariants in R.

THEOREM 3.6.1 Let L be a field, G a finite subgroup of Aut L, and K the set of G-invariants in L. Then K is a field and L is a Galois extension of K such that (1) G(L/K) = G and (2) #(G) = [L : K].

PROOF: K is obviously a field. Let a be an arbitrary element of L. Set $H_a = \{\sigma \in G \mid a^\sigma = a\}$. H_a is a subgroup and $a^\sigma = a^\tau$ (σ, $\tau \in G$) if and only if $H_a\sigma = H_a\tau$. Take a set of representatives $\sigma_1, \ldots, \sigma_r$ of $H_a\backslash G$ [$r = \#(G/H_a)$], set $a_i = a^{\sigma_i}$, and let $c_1, \ldots, c_r$ be the elementary symmetric forms of $a_1, \ldots, a_r$ of degrees 1, ..., r, respectively. Then these c_i are in K, and therefore $f_a(X) = \Pi_{i=1}^r (X - a_i)$ is a polynomial over K which has a as a root and is separable. Therefore we see that each element a of L is separable over K and $[K(a) : K] \leq \#(G)$. In particular, L is separably algebraic over K. Therefore every finite extension of K contained in L is a simple extension by Theorem 3.4.1. Since it was shown that the degree of such extension is at most #(G), we see that $[L : K] \leq \#(G)$. Since G consists of automorphisms of L over K, we see that $\#(G) \leq [L : K]$ (Theroem 3.4.2), and we see that #(G) = [L : K]; hence $G = \mathrm{Aut}_K L$. Our observation on $f_a(X)$ above shows us that L is a normal extension of K and that L is separable. Therefore L is a Galois extension of K. QED

COROLLARY 3.6.2 Let L be a finite normal extension of a field K, L_s the separable closure of K in L, and L_i the set of purely inseparable elements in L over K. Then:

(1) L_i is the field of invariants of $\mathrm{Aut}_K L = G(L/K)$ and $L = L_s \vee L_i$.

(2) L is a Galois extension of L_i and $G(L/L_i) \cong G(L/K)$.

(3) $[L : L_i] = [L_s : K]$ and $[L : L_s] = [L_i : K]$.

PROOF: Let K' be the field of invariants of $G(L/K)$. If there is $a \in K'$ which is not in L_i, then there is a conjugate a' of a which is not equal to a. Then there is a K-injection σ of $K(a)$ in a large field Ω so that $a^\sigma = a'$. This σ is extended to L and $\sigma \in G(L/K)$ because L is normal over K. Thus $K' \subseteq L_i$. Conversely, if $a \in L_i$, then a is the unique conjugate of a, and $L_i \subseteq K'$. Thus $K' = L_i$. Let L' be $L_s \vee L_i$. Since L is purely inseparable over L_s, we see that L is purely inseparable over L'. Since L is separable over $K' = L_i$ (Theorem 3.6.1), we see that L is separable over L'. These imply that $L = L'$. (2) and the equality $[L : L_i] = [L_s : K]$ follow from the observation we have made and Theorems 3.5.2 and 3.6.1. Then we have $[L : L_s] = [L : K]/[L_s : K] = [L : K]/[L : L_i] = [L_i : K]$. QED

THEOREM 3.6.3 Let K, L, K' be fields such that $K \subseteq L \cap K'$.

(1) If L is a Galois extension of K of finite degree, then $K' \vee L$ is a Galois extension of K' and $G(K' \vee L/K') \cong G(L/K' \cap L))$; hence $[K' \vee L : K'] = [L : K' \cap L]$.

(2) If L is a separably algebraic extension of K of finite degree and if K' is a normal extension of K, then $K' \vee L$ is separably algebraic over K' and $[K' \vee L : K'] = [L : K' \cap L]$.

(3) If L is a purely inseparable extension of K of finite degree and if K' is a separably algebraic extension of K, then $K' \vee L$ is a purely inseparable extension of K' and $[K' \vee L : K'] = [L : K' \cap L] = [L : K]$.

PROOF: (1): Let a be such that $L = K(a)$ and let $a = a_1$, $a_2, \ldots, a_r$ be the roots of the minimal polynomial $f(X)$ for a over K. Since L is Galois over K, all a_i are in L, and therefore $a_i \in K'(a) = K' \vee L$. This shows that $K' \vee L$ is Galois over K'. The minimal polynomial $g(X)$ for a over K' is a factor of $f(X)$, and we may assume that $g(X) = (X - a_1)\cdots(X - a_s)$ $(s \leq r)$.

Then the coefficients of $g(X)$ are in L; hence they are in $K' \cap L$ and therefore $[K' \vee L : K'] = [L : K' \cap L]$. For each element σ of $\mathrm{Aut}_{K'}\, K' \vee L$, its restriction on L induces an automorphism of L over $K' \cap L$. Let G be the subgroup of $\mathrm{Aut}_{K' \cap L}\, L$ consisting of these restrictions. Then G is a homomorphic image of $\mathrm{Aut}_{K'}\, K' \vee L$, and the field of G-invariants in L is $K' \cap L$ as is readily seen. Therefore $G = \mathrm{Aut}_{K' \cap L}\, L$. Therefore $\#(G) = [L : K' \cap L]$ (Theorem 3.6.1) and $\mathrm{Aut}_{K'}\, K' \vee L \cong G$. Thus we see (1).

(2): Let a, $f(X)$ and $g(X)$ be as above. Coefficients of $g(X)$ are expressed as polynomials in $a_1, \ldots, a_s$; hence they are separable over K. Since they are contained in K' which is normal over K, we see that there is a finite Galois extension K^* of $K' \cap L$ such that $K^* \subseteq K'$ (hence $K^* \cap L = K' \cap L$) and such that $g(X) \in K^*[X]$. Then, by virtue of (1) above, we have $[L \vee K^* : L] = [K^* : K^* \cap L] = [K^* : K' \cap L]$. Therefore $s \geq [L \vee K^* : K^*] = [L \vee K^* : K' \cap L]/[K^* : K' \cap L] = [L : K' \cap L]$. Obviously $s \leq [L : K' \cap L]$ and we prove (2).

(3): If the characteristic p of K is 0, then obviously $L = K$, and we assume that $p \neq 0$. We take a sequence $K = L_0 \subset L_1 \subset \cdots \subset L_m = L$ of intermediate fields L_i such that $[L_{j+1} : L_j] = p$ for each $j = 0, \ldots, m - 1$. Setting $L_j' = K' \vee L_j$, we have only to show that $[L_{j+1}' : L_j'] = p$ for every j. Thus we may assume that $[L : K] = p$. In this case, we take an element a of L which is not in K. Then $L = K(a)$ and $a^p \in K$. Then $a^p \in K'$, and therefore $[K'(a) : K']$ is either p or 1. If it were 1, then $a \in K'$; hence a is separable and purely inseparable over K, which implies that $a \in K$, contradicting our assumption. Thus $[K'(a) : K'] = p = [L : K]$. QED

3.7 THE FUNDAMENTAL THEOREM OF GALOIS

The following is called the <u>fundamental theorem of Galois</u>.

THEOREM 3.7.1 Let L be a Galois extension of a field K of finite degree. Then, for every intermediate field M, L is a Galois extension of M. Furthermore:

(1) There is a one-one correspondence between the set I of intermediate fields M and the set S of subgroups H of the Galois group $G(L/K)$ in such a way that: M corresponds to H if and only if M is the field of invariants of H in L, or, if and only if $H = G(L/M)$.

(2) Denoting by $H(M)$ the subgroup corresponding to M under the correspondence given above, we have the following result for intermediate fields M, M' : (i) $M \subseteq M'$ if and only if $H(M) \supseteq H(M')$, (ii) $H(M \cap M') = H(M) \vee H(M')$, (iii) $H(M \vee M') = H(M) \cap H(M')$, and (iv) if $\sigma \in G(L/K)$, then $H(M^{\sigma}) = \sigma^{-1}H(M)\sigma$.

(3) An intermediate field M is a Galois extension of K if and only if $H(M)$ is a normal subgroup of $G(L/K)$. In this case, the mapping of $G(L/K)$ which maps each element σ to its restriction on M is a homomorphism of $G(L/K)$ onto $G(M/K)$ and the kernel of the mapping is $H(M)$. Hence $G(M/K) \cong G(L/K)/H(M)$.

PROOF: Conjugates of an element a over M are conjugates of a over K (but not conversely), and therefore the statement at the beginning is obvious.

(1): Set $H(M) = G(L/M)$ $(= \mathrm{Aut}_M L)$. Theorem 3.4.2 shows that $\#(H(M)) = [L : M]$. Therefore Theorem 3.6.1 implies that M is the field of $H(M)$-invariants in L. Conversely, if H is a subgroup of $G(L/K)$, then for the field M' of H-invariants, we have $H = G(L/M')$ by Theorem 3.6.1, and we prove (1).

(2): (i) is obvious. As for (ii), since $M \cap M' \subseteq M$, we have $H(M \cap M') \supseteq H(M)$. Similarly, $H(M \cap M') \supseteq H(M')$ and $H(M \cap M') \supseteq H(M) \vee H(M')$. Let M^* be the field of $(H(M) \vee H(M'))$-invariants. Then since $H(M^*) \supseteq H(M) \vee H(M')$ we see that $M^* \subseteq M \cap M$; hence $H(M^*) \supseteq H(M \cap M') \supseteq H(M) \vee H(M') = H(M^*)$, which proves (ii). (iii) is proved similarly. $\sigma \in H(M^{\sigma})$ if and only if $a^{\sigma\tau} = a^{\sigma}$, i.e., $a^{\sigma\tau\sigma^{-1}} = a$, for every $a \in M$, which proves (iv).

(3): M is a Galois extension of K if and only if $M^{\sigma} = M$ for every $\sigma \in G(L/K)$ by Theorem 3.5.3. Therefore (iv) above shows the first statemant of (3). In this case, since $M^{\sigma} = M$ for every $\sigma \in G(L/K)$, the restriction of σ on M is an element of $G(M/K)$. For two elements σ, τ of $G(L/K)$, (i) if $H(M)\sigma = H(M)\tau$, then since every element of H(M) fixes elements of M, we see that $\sigma|_M = \tau|_M$; conversely, if $\sigma|_M = \tau|_M$, then $H(M)\sigma\tau^{-1} = H(M)$ and $(\sigma\tau^{-1})|_M$ is the identity, which implies that $\sigma\tau^{-1} \in H(M)$. Thus the image of G(L/K) by the restriction map is isomorphic to G(L/K)/H(M). #(G(L/K)/H(M)) = #(G(L/K)/#(H(M)) = [L : K]/[L : M] = [M : K] = #(G(M/K)), and we complete the proof. QED

3.8 ROOTS OF UNITY AND CYCLIC EXTENSIONS

Let L be a splitting field for $X^n - 1$ over a field K of characteristic p (which may be zero), where n is a given natural number. Then:

THEOREM 3.8.1 The set U of n-th roots of unity (i.e., the roots of $X^n - 1$) in L forms a cyclic group. (1) If p = 0, then #(U) = n. (2) If p ≠ 0 and if n is not divisible by p, then #(U) = n. (3) If p ≠ 0 and if $n = p^e m$ (e is a natural number and m is a natural number which is not divisible by p), then #(U) = m and U coincides with the set of m-th roots of unity.

In cases (1) and (2), a generator of U is called a primitive n-th root of unity.

PROOF; (1) and (2): $\frac{d}{dX}(X^n - 1) = nX^{n-1}$ and this has no common root with $X^n - 1$. Therefore all roots of $X^n - 1$ are simple roots. Therefore #(U) = n. Obviously U forms a group. Since $\#(\{a \in U \mid a^s = 1\}) \leq \#(\{a \in L \mid a^s = 1\}) \leq s$ for every natural number s, we see that U is cyclic (cf. the proof of Theorem 3.3.7).

(3): In this case, 1 is the unique solution for $X^{p^e} = 1$ [because $X^{p^e} - 1 = (X - 1)^{p^e}$]. Therefore $a^n = 1$ if and only if $a^m = 1$. QED

Let L be a Galois extension of a field K of finite degree. If G(L/K) is commutative, then we say that L is an abelian extension of K. If G(L/K) is cyclic, then we say that L is a cyclic extension of K. A minimal splitting field for $X^n - 1$ over a field K is called the cyclotomic field of order n over K. A field L is called a cyclotomic field over K if it is an intermediate field between K and a cyclotomic field of certain order over K.

THEOREM 3.8.2 A cyclotomic field L over a field K is an abelian extension.

PROOF: L is a subfield of the minimal splitting field L* for $X^n - 1$, where we may assume that $X^n - 1$ is separable by virtue of Theorem 3.8.1. Let ζ be a primitive n-th root of unity. Then L* = $K(\zeta)$. If $\sigma \in G(L^*/K)$, then $\zeta^\sigma = \zeta^i$ $(1 \leq i < n$; i has no proper common factor with n). Since such an i defines σ uniquely, we see that $\sigma \rightsquigarrow i$ induces a one-one correspondence between elements σ of $G(L^*/K)$ and some units in the ring $\mathbb{Z}/n\mathbb{Z}$. If $\sigma \rightsquigarrow i$, $\tau \rightsquigarrow j$, then $\zeta^{\sigma\tau} = (\zeta^i)^\tau = (\zeta^j)^i = \zeta^{ij}$. Thus we have an injection of $G(L^*/K)$ into the unit group of $\mathbb{Z}/n\mathbb{Z}$. Therefore $G(L^*/K)$ is commutative. By Theorem 3.7.1, L corresponds to a subgroup H of $G(L^*/K)$ and, in view of the fact that $G(L^*/K)$ is commutative, we see that L is a Galios extension of K and that $G(L/K) = G(L^*/K)/H$. Thus L is an abelian extension of K. QED

THEOREM 3.8.3 If L is an algebraic extension of finite degree over a finite field K, then L is a cyclic extension of K.

PROOF: Set $q = \#(L)$. Then L is the minimal splitting field for $X^{q-1} - 1$ (cf. the footnote to Theorem 3.1.2), and $G(L/K)$ is therefore commutative as proved above. Let d be any factor of $[L : K] = \#(G(L/K))$, and assume that H and H' are subgroups of order d of $G(L/K)$. Then the fields M, M' of H-invariants and H'-invariants, respectively, in L consist of the same number of elements, say s (because $d = [L : M] = [L : M']$). Then both M and M' are the minimal splitting fields for $X^{s-1} - 1$; hence $M = M'$. This shows that $G(L/K)$ is cyclic by virtue of Lemma 1.6.1. QED

COROLLARY 3.8.4 A cyclotomic field over a field of nonzero characteristic is a cyclic extension.

PROOF: This follows from Theorems 3.6.3 and 3.8.3. QED

Let P_n be the cyclotomic field of order n over the rational number field Q and we are going to prove the following theorem.

THEOREM 3.8.5 $G(P_n/\mathbb{Q})$ is isomorphic to the unit group U_n of $\mathbb{Z}/n\mathbb{Z}$.

In order to prove this, we observe the n-th cyclotomic polynomial $\kappa_n(X)$ defined to be $\Pi_\zeta (X - \zeta)$, where ζ runs through all primitive n-th roots of unity. Note that if ζ is one of these ζ, then others are ζ^i $[1 < i < n;\ (i \bmod n) \in U_n]$ (cf. Theorem 3.8.1). If we denote by $\phi(n)$ the degree of $\kappa_n(X)$, then ϕ is an integer-valued function defined on the set of natural numbers. This function is called the Euler function. Now we state:

THEOREM 3.8.6 Over the rational number field Q, the cyclotomic polynomial $\kappa_n(X)$ is in $Z[X]$ (Z is the ring of rational integers) and is irreducible.

First we prove the following lemma.

LEMMA 3.8.7 If K is a finite field, if $q = \#(K)$, and if $f(X) \in K[X]$, then $f(X)^q = f(X^q)$.

PROOF: Write $f(X)$ as $\sum c_i X^i$. Then since $c_i^q = c_i$ (Theorem 3.1.2), we have $f(X)^q = \sum c_i^q X^{iq} = \sum c_i X^{iq} = f(X^q)$ (Lemma 3.1.3). QED

PROOF of THEOREM 3.8.6: Obviously, if $\sigma \in G(P_n/Q)$, then $(\kappa_n(X))^\sigma = \kappa_n(X)$ and therefore $\kappa_n(X) \in Q[X]$. Furthermore, all roots are integral over Z. Therefore the coefficients of $\kappa_n(X)$ are integral over Z (Corollary 2.9.3), and hence $\kappa_n(X) \in Z[X]$ (Theorem 2.9.5). Since $Z[X]$ is a UFD, if $\kappa_n(X)$ were reducible in $Q[X]$, it is reducible in $Z[X]$ by Lemma 2.6.5. So, we assume that $\kappa_n(X) = h(X)k(X)$ with an irreducible monic polynomial $h(X)$ and a $k(X)$. Let H be the set of roots of $h(X)$ and let $\zeta \in H$. Take a root of $\kappa_n(X)$ which is not in H. It is of the form ζ^s $[1 < s < n,\ (s \bmod n) \in U_n]$. Choose such a root so that s is smallest. Then, taking a prime factor q of s, we see that $\zeta^{s/q} \in H$. Therefore, choosing $\zeta^{s/q}$ instead of ζ, we may assume that $\zeta^q \notin H$, i.e., that ζ^q is a root of $k(X)$. This means that $h(X)$ and $k(X^q)$ have a common root ζ. Then $h(X)$ and $k(X^q)$ have a nontrivial common factor (Proposition 2.6.7). Since $h(X)$ is irreducible, it follows that $h(X)$ is a factor of $k(X^q)$ and $k(X^q) = h(X)g(X)$. Let us denote by * the natural homomorphism $Z \to Z/qZ$. Then, by virtue of Lemma

3.8.7, we have $k^*(X)^q = k^*(X^q) = h^*(X)g^*(X)$. Therefore $k^*(X)$ and $h^*(X)$ must have a nontrivial common factor, which implies that $\kappa_n^*(X)$ has a multiple root. But, since $\kappa_n(X)$ is a factor of $X^n - 1$ and since q does not divide n by our choice of q, we see that $\kappa_n^*(X)$ cannot have any multiple root. Thus $\kappa_n(X)$ must be irreducible. QED

PROOF of THEOREM 3.8.5: As we saw in the proof of Theorem 3.8.2, $G(P_n/Q)$ is isomorphic to a subgroup of U_n. The irreducibility of $\kappa_n(X)$ shows that if $(s \bmod n) \in U_n$ and if ζ is a primitive n-th root of unity, then there is a σ in $G(P_n/Q)$ which maps ζ to ζ^s, and we see that $G(P_n/Q) = U_n$. QED

In the rest of this section, we aim to characterize cyclic extensions in two important cases. The first case is:

THEOREM 3.8.8 Let K be a field of characteristic p (which may be zero), and let n be a natural number which is not a multiple of p. Assume that K is splitting field for $X^n - 1$ (this is often expressed by saying that K contains all n-th roots of unity). Let L be an algebraic extension of K of degree n. Then L is a cyclic extension of K if and only if there is an element b of L such that $L = K(b)$ and such that $b^n \in K$.

NOTATION: (1) If a is an element of a field K and if $X^n - a$ is irreducible over K, then one of the roots of $X^n - a$ is denoted by $\sqrt[n]{a}$ or by $a^{1/n}$. Under the notation, the conclusion of the theorem can be expressed by $L = K(a^{1/n})$ with a suitable $a \in K$. (2) If c is an element of a field L and $\sigma_1, \ldots, \sigma_m$ are elements of Aut L, we mean by $c^{\sigma_1+\cdots+\sigma_m}$ the element $c^{\sigma_1}\cdots c^{\sigma_m}$. (3) More generally, if $\tau_1, \ldots, \tau_m$ are injections of a field L into a field Ω and if $b_1, \ldots, b_m$ are elements of Ω, then we have a module homomorphism of L in Ω defined by $x \rightsquigarrow \sum b_i x^{\tau_i}$. This module homomorphism in denoted by $\sum \tau_i b_i$.

PROOF of THEOREM 3.8.8: The if part is left to the reader (Exercise 3.8.1) and we consider the only if part. Letting σ be a generator of $G(L/K)$, we consider <u>Lagrange resolvent</u> for two elements x, t of L defined as follows:

$$u(d, t) = t + xt^{\sigma} + x^{1+\sigma}t^{\sigma^2} + x^{1+\sigma+\sigma^2}t^{\sigma^3} + \cdots + x^{1+\sigma+\cdots+\sigma^{n-2}}t^{\sigma^{n-1}}$$

Then, as is easily seen, if $x^{1+\sigma+\cdots+\sigma^{n-1}} = 1$, it holds that $x(u(x,t))^{\sigma} = u(x,t)$. Now let ζ be a primitive n-th root of unity and assume the existence of t such that $u(\zeta,t) \neq 0$. Set $b = u(\zeta,t)$. Then $b \in L$. Since $\zeta \in K$, we have $\zeta^{\sigma} = \zeta$ and $\zeta^{1+\cdots+\sigma^{n-1}} = 1$. Hence $b^{\sigma} = \zeta^{-1}b$. Thus the set of conjugates of b are $\zeta^{i}b$ $(i = 0, 1, \ldots, n - 1)$. Therefore this is the required element. The existence of t assumed above follows from:

LEMMA 3.8.9 Under the notation (3) above, if $\tau_1, \ldots, \tau_m$ are distinct from each other and if some of b_i are not zero, then the mapping $\sum \tau_i b_i$ is different from zero.

PROOF: Assume the contrary. Omitting some of the τ_i if necessary, we may assume that $\sum \tau_i b_i = 0$, with b_i all different from 0. Obviously $m \geq 2$. Since $\tau_1 \neq \tau_2$, there is u in L such that $u^{\tau_1} \neq u^{\tau_2}$. We apply the mapping to ux $(x \in L)$ and we see that $0 = \sum b_i u^{\tau_i} x^{\tau_i}$. Therefore we see that $\sum \tau_i (b_i u^{\tau_i}) = 0$. Hence we see that $0 = (\sum \tau_i b_i)u^{\tau_1} - \sum \tau_i (bu^{\tau_i}) = \tau_2(u^{\tau_1} - u^{\tau_2})b_2 + \cdots + \tau_m(u^{\tau_1} - u^{\tau_m})b_m$. Thus we obtained a case with smaller number of τ_i, and we complete the proof by induction on m. QED

The other case we want to deal with is the case of a cyclic extension of degree p = the characteristic of the field.

THEOREM 3.8.10 Let K be a field of characteristic $p \neq 0$.

(1) If L is a cyclic extension of degree p over K, then there is c in K such that $L = K(b)$ with a root b of $X^p - X - c$.

(2) Conversely, if $c \in K$ and if b is a root of $f_c(X) = X^p - X - c$, then (i) roots of $f_c(X)$ are $b, b + 1, \ldots, b + (p-1)$; and (ii) if $b \notin K$, then $K(b)$ is a cyclic extension of degree p over K.

PROOF: If b is a root of $f_c(X)$ and if n is an element of the prime field P, then $(b + n)^p = b^p + n^p = b + c + n$ and therefore $b + n$ is also a root of $f_c(X)$, which proves (i) in (2). Assume that $b \notin K$. Then there is a conjugate $b + n$ of b with $0 \neq n \in P$. Let σ be an element of the Galois group of the minimal splitting field for $f_c(X)$ over K such that $b^\sigma = b + n$. Then $b^{\sigma^i} = b + ni$ are all conjugates of b. Thus b, $b + 1$, ..., $b + (p - 1)$ are conjugates of b. Thus $K(b)$ is a Galois extension of K, $p = [K(b) : K]$, and we have proved (ii) in (2). We assume now that L is a cyclic extension of degree p over K. Let σ be a generator of $G(L/K)$. By Lemma 3.8.9, the module homomorphism $\sum_{i=0}^{p-1} \sigma^i$ of L into L is different from zero. This means that there is u in L such that $\sum u^{\sigma^i} \neq 0$. Set $v = u^\sigma + 2u^{\sigma^2} + \cdots + (p - 1)u^{\sigma^{p-1}}$. Then $v - v^\sigma = u^\sigma + u^{\sigma^2} + \cdots + u^{\sigma^{p-1}} - (p - 1)u = u^\sigma + u^{\sigma^2} + \cdots + u^{\sigma^{p-1}} + u$. This is an element of K, and by our choice of u, this is not zero. Set $d = v - v^\sigma$. We see that $v^{\sigma^i} = v - id$ because $v^\sigma = v - d$. Therefore the conjugates of v are v, $v - d$, $v - 2d$, ..., $v - (p - 1)d$. Set $b = v/d$. Then the conjugates of b are b, $b - 1$, ..., $b - (p - 1)$, and $L = K(b)$. Set $c = b(b - 1)\cdots(b - (p - 1))$. Since $X^p - X = \Pi_{a \in P} (X - a)$, we see that $b^p - b = c$. Thus we see that $f_c(X) = X^p - X - c$ is the minimal polynomial for b over K. QED

3.9 SOLVABILITY OF ALGEBRAIC EQUATIONS BY RADICALS

An algebraic extension field L of finite degree over a field K is called an <u>extension by radicals</u> if there is a sequence of intermediate fields L_i such that $K = L_0 \subset L_1 \subset \cdots \subset L_r = L$ and such that each L_i $(i \geq 1)$ is $L_{i-1}(b_i)$ with $b_i = a_i^{1/m_i}$, $a_i \in L_{i-1}$ (hence, by our assumption, $X^{m_i} - a_i$ is irreducible over L_{i-1}). An element c of an extension field of K is said to be <u>expressible by radicals</u> over K if it is contained in an extension by radicals. An equation $X^n + c_1X^{n-1} + \cdots + c_n = 0$ over a field K is said to be <u>solvable by radicals</u> if every root is expressible by radicals over the field generated by the coefficients $c_1, \ldots, c_n$ (over the prime field).

The following lemma is obvious:

LEMMA 3.9.1 For fields K, L, M, if M and L are extensions by radicals of L and K, respectively, then M is an extension by radicals of K.

THEOREM 3.9.2 Over a field K of characteristic zero, every root of unity is expressible by radicals. For each natural number n, let U_n be the set $\{\zeta \mid \zeta^m = 1, m \leq n\}$. Then the field K_n generated by U_n over K is an extension by radicals.

PROOF: We prove the last statement by induction on n. $K_1 = K_2 = K$, and the assertion is obvious for $n \leq 2$. Assume that $n \geq 3$. K_n is the minimal splitting field for $X^n - 1$ over K_{n-1}. Since $X^n - 1$ is divisible by $X - 1$, and since K_n is an abelian extension of K_{n-1} by Theorem 3.8.2, we see that $G(K_n/K_{n-1})$ is a commutative group of order at most $n - 1$. Take a composition series $G(K_n/K_{n-1}) = G_0 \supset G_1 \supset \cdots \supset G_s = \{1\}$ and consider the corresponding sequence of intermediate fields: $K_{n-1} = L_0 \subset L_1 \subset \cdots \subset L_s = K_n$ (Theorem 3.7.1). Then Theorem 3.8.8 shows that each L_i is an extension by radicals of L_{i-1}. Therefore we complete the proof by Lemma 3.9.1. QED

LEMMA 3.9.3 Assume that $X^n - a$ is irreducible over a field K $(a \in K)$ and let ζ be a primitive n-th root of unity. If a field K^* contains $K(\zeta)$, then the minimal polynomial for $a^{1/n}$ over K^* is $X^s - b$ with a divisor s of n and a suitable element b of K^*.

PROOF: Conjugates of $a^{1/n}$ over K^* are roots of $X^n - a$; hence they are of the form $\zeta^i a^{1/n}$. Let them be $\zeta^i a^{1/n}$ with $i = m_1, \ldots, m_s$. Their product coincides with $\zeta^t (a^{1/n})^s$ $(t = \sum m_i)$. Set $b = (a^{1/n})^s$. Since $\zeta \in K^*$, we have $b \in K^*$. Thus $X^s - b$ is a polynomial over K^* and has $a^{1/n}$ as a root. Since $a^{1/n}$ has s conjugates, $X^s - b$ must be irreducible over K^*, and hence must be the minimal polynomial for $a^{1/n}$ over K^*. The roots of $X^s - b$ are simple roots because of our observation. Therefore they must be $\eta^i a^{1/n}$ $(i = 0, 1, \ldots, s - 1)$, where η is a primitive s-th root of unity. On the other hand, conjugates of $a^{1/n}$ are $\zeta^i a^{1/n}$ $(i = m_1, \ldots, m_s)$. Therefore some of ζ^{m_i} must be a primitive s-th root of unity, which shows that s is a divisor of n. QED

COROLLARY 3.9.4 If a field L is an extension by radicals over a field K of characteristic 0 and if K' is a field containing K, then $K' \vee L$ is contained in an extension by radicals over K'.

PROOF: Take a sequence $K = L_0 \subset L_1 \subset \cdots \subset L_r = L$ such that $L_i = L_{i-1}(a_i^{1/n_i})$ with $a_i \in L_{i-1}$. Let m be a common multiple of $n_1, \ldots, n_r$ and consider K_m in Theorem 3.9.2. Then $K_m \vee L \vee K'$ is an extension by radicals over K'. QED

Now we prove the following:

THEOREM 3.9.5 Assume that L is a Galois extension of a field K of characteristic zero. Then L is contained in an extension by radicals over K if and only if the Galois group $G(L/K)$ is solvable.

PROOF: Assume that $G(L/K)$ is solvable, and let $G(L/K) = G_0 \supset G_1 \supset \cdots \supset G_r = \{1\}$ be a composition series. Let m be a common multiple of $\#(G_{i-1}/G_i)$ $(i = 1, \ldots, r)$. Consider K_m in Theorem 3.9.2. K_m is an extension by radicals over K. $L^* = L \vee K_m$ is a Galois extension of K_m whose Galois group is isomorphis to $G(L/(L^* \cap K_m))$ (Theorem 3.6.3), which is solvable. Then applying our proof of Theorem 3.9.2 to this case, we see that L^* is an extension by radicals over K_m, and hence over K, too. Conversely, assume that L is a subfield of an extension by radicals over K, namely that there is a sequence $K = L_0 \subset L_1 \subset \cdots \subset L_r$ such that $L \subseteq L_r$ and $L_i = L_{i-1}(a_i^{1/n_i})$ $(i = 1, \ldots, r)$. Let m be a common multiple of $n_1, \ldots, n_r$ and consider K_m in Theorem 3.9.2. Then, considering $K_m \vee L_r$ instead of L_r, we may assume that (i) some L_i, say L_c, coincides with K_m and (ii) m is a multiple of n_i for $i > c$. Then, in view of Lemma 3.9.3, we see that the field generated by conjugates of L_r over K is an extension by radicals having similar sequence, and we may assume that L_r is a Galois extension of K. Now, $G(L_c/K)$ is commutative by Theorem 3.8.2. As for $i > c$, $G(L_i/L_{i-1})$ is cyclic by Theorem 3.8.8, and we see that $G(L_r/K)$ is solvable. $G(L/K) \cong$

$G(L_r/K)/G(L_r/L)$ (Theorem 3.7.1) and therefore $G(L/K)$ is solvable. QED

COROLLARY 3.9.6 Let $f(X) = X^n + c_1X^{n-1} + \cdots + c_n$ be a polynomial with coefficients c_i in a field of characteristic zero. Then $f(X) = 0$ is solvable by radicals if and only if the Galois group of $f(X)$ is solvable.

This fact is acturally the origin of the term "solvable" groups.

In connection with the result given above, we give some comments on the known methods solving algebraic equations of degree at most 4 and the nonsolvability in higher degree cases.

In general, for polynomial $f(X) = a_0X^n + a_1X^{n-1} + \cdots + a_n$ with coefficients a_i in a field K ($a_0 \neq 0$), the discriminant D of $f(X)$ [or of $f(X) = 0$] is defined to be the square of the difference product Δ of the roots of $f(X)$; i.e., denoting by $\alpha_1, \ldots, \alpha_n$ the roots of $f(X)$, we have $D = \Pi_{i<j}(\alpha_i - \alpha_j)^2$. Since the difference product is an alternating form, its square is a symmetric form. Since the elementary symmetric forms of roots of $f(X)$ are $-a_1/a_0, a_2/a_0, \ldots, (-1)^n a_n/a_0$, we see that D is expressed as a polynomial in $c_1 = a_1/a_0, \ldots, c_n = a_n/a_0$ by Theorem 2.8.1. Obviously, the difference product Δ is in the minimal splitting field L_f of $f(X)$ over the field K_f generated by $c_1, \ldots, c_n$, and we have the following lemma.

LEMMA 3.9.7 With the notation above, D is in K_f and $X^2 - D$ has roots $\pm\Delta$ in L_f.

Another remark is that if σ is an element of the Galois group G of $f(X)$, then σ defines a permutation of roots of $f(X)$, and this gives an injection of G into the symmetric group on the set of roots of $f(X)$. Thus we have:

LEMMA 3.9.8 The Galois group G of $f(X)$ is a permutation group on the set of roots of $f(X)$.

Let us begin with the case of quadratic equation, i.e., $n = 2$. Hereafter, we assume for simplicity that the characteristic is zero.

In this case, unless $L_f = K_f$, $\Delta \notin K_f$ and $L_f = K_f(\Delta)$. Therefore $f(X) = 0$ is solved substantially by solving $X^2 - D = 0$.

Next, we observe the cubic case, i.e., $n = 3$. Then the symmetric group S_3 has order 6. Therefore $\#(G)$ is one of 1, 2, 3, 6. Note also that permutations fixing the difference product are elements of the alternating group A_3 and $\#(A_3) = 3$. Therefore $[L_f : K_f(\Delta)]$ is either 1 or 3. If it is 1, then $f = 0$ is substantially solved by solving $X^2 - D = 0$. If it is 3, then we (i) solve $X^2 - D = 0$ and obtain $\pm\Delta$ and (ii) consider $\omega = (-1 + \sqrt{-3})/2$, a cubic root of unity. (iii) Then over $K_f(\Delta,\omega)$, L_f generates a cyclic extension of degree 3 which is generated by $\alpha^{1/3}$ $(\alpha \in K_f(\Delta,\omega))$ by Theorem 3.8.8.

Now we observe the case $n = 4$. The alternating group A_4 on $\{1,\ldots,4\}$ has a normal subgroup N consisting of elements 1, $(1,2)\cdot(3,4)$, $(1,3)(2,4)$, $(1,4)(2,3)$. (i) $G \cap A_4$ corresponds to $K_f(\Delta)$ as in the cubic case. (ii) $(G \cap A_4)/(G \cap N)$ is a cyclic group of order 1 or 3 (because A_4/N is a cyclic group of order 3), and therefore the step of finding out a field corresponding to $G \cap N$ is similar to the cubic case. Hence, in the known method of solving a quartic equation, we solve a suitable cubic equation in this step. (iii) The last step has Galois group $G \cap N$ which is either the direct product of two cyclic groups of order 2 or its subgroup. Hence the step is reduced to solve at most two quadratic equations.

Last we observe the higher degree case. For our purpose, we admit the following fact for a moment (the proof will be given at the end of Sec. 3.11).

LEMMA 3.9.9 There are real numbers $b_1, \ldots, b_n$ which are algebraically independent over the rational number field Q.

Then we have:

THEOREM 3.9.10 For every natural number n, there is a polynomial $f_n(X) = X^n + c_1X^{n-1} + \cdots + c_n$ (c_i are real numbrs) such that the Galois group G of $f_n(X)$ is isomorphic to the symmetric group S_n of degree n.

PROOF: Let $b_1, \ldots, b_n$ be as in Lemma 3.9.9 and consider $f_n(X)$

$= \prod_i (X - b_i)$. Then $f_n(X) = X^n + c_1X^{n-1} + \cdots + c_n$ with c_i such that $(-1)^i c_i$ $(i = 1, \ldots, n)$ are the elementary symmetric forms of degree i in $b_1, \ldots, b_n$. The symmetric group S_n acts on $\{b_1,\ldots,b_n\}$ naturally, and hence to $Q(b_1,\ldots,b_n)$. Let K be the field of S_n-invariants. Since c_i are symmetric forms, K contains $Q(c_1,\ldots,c_n)$. Let k be an arbitrary element of K. Then $k = g(b_1, \ldots,b_n)/h(b_1,\ldots,b_n)$ with polynomials g, h in $b_1, \ldots, b_n$ with coefficients in Q, and such that g and h have no proper common factor. Let σ be an arbitrary element of S_n. Since $g^\sigma/h^\sigma = k^\sigma = k = g/h$, we have $g^\sigma h = h^\sigma g$. Since g and h have no common factor, it follows that g divides g^σ and h divides h^σ. Since degrees are invariant under σ, we see that there is an element e_σ of Q such that $g^\sigma = e_\sigma g$, $h^\sigma = e_\sigma h$. For $\sigma,\tau \in S_n$, we have $g^{\sigma\tau} = (e_\sigma g)^\tau = e_\sigma e_\tau g$ and $e_{\sigma\tau} = e_\sigma e_\tau$. This means that $\sigma \leadsto e_\sigma$ gives a homomorphism of S_n in the multiplicative group Q^* of Q. In particular, if τ is a transposition, then $\tau^2 = 1$, whence $e_\tau^2 = 1$. Therefore g and h are either symmetric forms or alternating forms. If they are alternating forms, then they are divided by the difference product of b_i (Theorem 2.8.2), which contradicts our assumption that g, h have no common factor. Thus g, h are symmetric forms and $k = g/h \in Q(c_1,\ldots,c_n)$. Thus $K = Q(c_1,\ldots,c_n)$. Therefore the Galois group of $f_n(X)$ is the symmetric group S_n. QED

Though we do not prove it here, if $n \geq 5$, then S_n is not solvable (actually A_n is noncommutative and has no proper normal subgroup). Therefore $f_n(X) = 0$ is not solvable by radicals.

3.10 PROBLEM OF GEOMETRIC CONSTRUCTION

The problem of geometric construction is very classical and discussions were held even in ancient Greece on such problems as (i) trisection of an angle, (ii) duplication of a cube (i.e., constructing $2^{1/3}$), and (iii) quadrature of a circle (i.e., constructing the square root of the ratio π of the circumference of a circle to the diameter). These three have been known to be impossible since the last century.

We now discuss a criterion for the possibility of a geometric construction and then discuss cases (i) to (iii) because they are historically significant.

In our discussion, a geometric construction is a plane construction using rule and compass only. A rule is used only to draw a line through two given or constructed points. A compass is used to make a circle whose center is a given or consructed point and whose radius is either not specially known or of length which is given or constructed. If no special point or length is given in a problem, we understand that two points are given whose distance is the unit length.

On the other hand, a line or a segment is given by two points, a circle is given by its center and radius, a length is given by a segment (hence by two points), and an angle is given by its vertex and edges of length one (hence by three points). Thus in any construction problem, given conditions are expressed by points, and the objects to be constructed are also expressed by points.

In expressing a point on a plane, we can use coordinates, say (a,b), where a, b are real numbers. Then we can associate a complex number $a + b\sqrt{-1}$. Thus, considering the so-called Gauss plane, we express points on a plane by complex numbers.

In the following (only in this section), Greek letters denote complex numbers, overbarred Greek letters denote complex conjugates.

THEOREM 3.10.1 When 0, 1, $\alpha_1, \ldots, \alpha_n$ are given points on a plane, we can obtain a point β by a geometric construction starting with them if and only if there is a sequence of fields $L_0 \subset L_1 \subset \cdots \subset L_s$ such that (i) $L_0 = Q(\alpha_1,\ldots,\alpha_n, \bar{\alpha}_1,\ldots,\bar{\alpha}_n)$, (ii) $[L_i : L_{i-1}] = 2$ for each $i = 1, \ldots, s$, and (iii) $\beta \in L_s$.

PROOF: The if part: The point of symmetry of a point with respect to a given line can be constructed, and therefore $\bar{\alpha}_i$ are constructed. If two complex numbers α, α' ($\neq 0$) are given, then (i) the product $\alpha\alpha'$ and the ratio α/α' can be constructed because addition and subtraction of arguments and the product and ratio of absolute values of α and α' can be constructed and (ii) $\alpha \pm \alpha'$ can be constructed obviously. Thus elements of L_0 are constructed.

If α is given, the $\sqrt{\alpha}$ is also constructed because half of a given angle and $\sqrt{|\alpha|}$ are constructed. Therefore if every element of L_{i-1} is constructed, then every element of L_i is also constructed. Thus every element of L_s is constructed. This proves the if part. The only if part: Let us consider those cases where we construct a new point from given points: (i) the intersection of two lines, (ii) one of the intersections of a line with a circle, and (iii) one of the intersections of two circles.

Case (i): Assume that γ_1, γ_2, γ_3, γ_4 are known points, $\gamma_1 \neq \gamma_2$, $\gamma_3 \neq \gamma_4$ and the new point δ is the intersection of the lines ℓ_1 and ℓ_2 going through γ_1, γ_2 and γ_3, γ_4, respectively. Points on ℓ_1 are expressed as $\gamma_1 + t(\gamma_2 - \gamma_1)$ with real numbers t. Points on ℓ_2 are $\gamma_3 + u(\gamma_4 - \gamma_3)$ with real numbers u. Hence δ is the solution of $\gamma_1 + t(\gamma_2 - \gamma_1) = \gamma_3 + u(\gamma_4 - \gamma_3)$. Since t, u are real numbers, it follows that $\overline{\gamma}_1 + t(\overline{\gamma}_2 - \overline{\gamma}_1) = \overline{\gamma}_3 + u(\overline{\gamma}_4 - \overline{\gamma}_3)$. If $(\gamma_2 - \gamma_1)(\overline{\gamma}_4 - \overline{\gamma}_3) - (\overline{\gamma}_2 - \overline{\gamma}_1)(\gamma_4 - \gamma_3) = 0$, it follows that $(\gamma_2 - \gamma_1)(\overline{\gamma}_4 - \overline{\gamma}_3)$ is a real number, which means that the difference of arguments of $\gamma_2 - \gamma_1$ and $\gamma_4 - \gamma_3$ is a multiple of π ($= 180°$); hence these lines are parallel, which is not the case. Therefore t is rationally expressed by $\gamma_1, \ldots, \gamma_4, \overline{\gamma}_1, \ldots, \overline{\gamma}_4$, and hence $\delta \in Q(\gamma_1,\ldots,\gamma_4,\overline{\gamma}_1,\ldots,\overline{\gamma}_4)$.

Case (ii): Assume that a line ℓ goes through α, γ $(\alpha \neq \gamma)$ and the new point δ is one of the intersections of ℓ and a cirle with center η and radius r. Then $\delta = \alpha + t(\gamma - \alpha)$ with a real number t and $|\delta - \eta| = r$. Therefore $(\alpha + t(\gamma - \alpha) - \eta)(\overline{\alpha} + t(\overline{\gamma} - \overline{\alpha}) - \overline{\eta}) = r^2$. Thus t is a root of a quadratic equation. Therefore δ is an element of a quadratic extension of $Q(\alpha, \gamma, \eta, \overline{\alpha}, \overline{\gamma}, \overline{\eta})$.

Case (iii): Assume that the new point δ is one of the intersections of two circles with centers η_1, η_2 and radii r_1, r_2, respectively $(\eta_1 \neq \eta_2)$. Then $(\delta - \eta_i)(\overline{\delta} - \overline{\eta}_i) = r_i^2$ $(i = 1, 2)$, and $\delta\overline{\delta} - (\overline{\delta}\eta_i + \delta\overline{\eta}_i) + \eta_i\overline{\eta}_i = r_i^2$. Taking the difference, we have $(\eta_2 - \eta_1)\overline{\delta} + (\overline{\eta}_2 - \overline{\eta}_1)\delta + \eta_1\overline{\eta}_1 - \eta_2\overline{\eta}_2 = r_1^2 - r_2^2$. Since $\eta_1 \neq \eta_2$, we can solve the last equation in $\overline{\delta}$ using δ. Then, from the first equation, we have a quadratic equation on δ over $Q(\eta_1,\eta_2,\overline{\eta}_1,\overline{\eta}_2,r_1^2, r_2^2)$.

In any case, the new point lies in the field of known points or in its quadratic extension. Therefore we complete the proof of the only if part. QED

COROLLARY 3.10.2 In the above theorem, the condition is equivalent to the existence of a similar sequence with the additional condition that L_s is a Galois extension of L_0.

PROOF: Assume that a sequence as in Theorem 3.10.1 exists. Then conjugates of L_s are obtained by successive quadratic extensions of L_0; the field generated by all the conjugates of L_s is obtained similarly. The converse is obvious. QED

COROLLARY 3.10.3 Under the circumstances of Theorem 3.10.1, if β is obtained by geometric construction, the degree of the minimal polynomial for β over L_0 must be a power of 2. (But not conversely; cf. Exercise 3.10.2.)

As applications of these results, we discuss three problems which were stated at the beginning of this section.

As for the quadrature of a circle, impossiblility is known by the fact that the ratio π is transcendental over Q (we do not prove the transcendence property of π).

Impossibility of the duplication of a cube follows from Corollary 3.10.3, because $X^3 - 2$ is irreducible over Q.

A proof of the impossibility of trisection of an angle can be given easily by showing that there is a complex number α such that $|\alpha| = 1$ and such that $X^3 - \alpha$ is irreducible over $Q(\alpha)$. In connection with this, we want to discuss the possibility of geometric constructions of regular polygons. This is equivalent to the possibility of geometric constructions of angles $2\pi/n$ when only the unit length is given. Thus it is equivalent to the possibility for primitive n-th roots of unity. Note that the cyclotimic polynomial $\kappa_n(X)$ of order n is irreducible over Q and its degree $\phi(n)$ is equal to $\#(U_n)$, where U_n is the set of natural numbers m such that $1 \leq m < n$ and such that m and n have no proper common divisor (Theorem 3.8.6). Therefore, if n is a prime number, then $\phi(n) = n - 1$.

THEOREM 3.10.4 If p is a prime number, then a regular p-gon can be obtained by a geometric construction if and only if $p - 1$ is a power of 2.

PROOF: The only if part follows from Corollary 3.10.3 in view of the fact that $\phi(p) = p - 1$. Conversely, if $p - 1 = 2^s$ with a natural number s, then denoting by ζ a primitive p-th root of unity, we see that $Q(\zeta)$ is an abelian extension of Q (Theorem 3.8.2) whose Galois group G has order 2^s. Hence G has a composition series $G = G_0 \supset G_1 \supset \cdots \supset G_s = \{1\}$ with $\#(G_{i-1}/G_i) = 2$ for $i = 1, \ldots, s$. Hence, by Theorem 3.10.1, we can obtain ζ by a geometric construction. QED

As for a general regular polygon, we assert the following:

THEOREM 3.10.5 Let n be a natural number ≥ 3. Then we can obtain a regular n-gon by a geometric construction if and only if $n = 2^s p_1 \cdots p_t$ with nonnegative rational integer s and mutually distinct odd prime numbers $p_1, \ldots, p_t$ such that $p_i - 1$ are all powers of 2.

PROOF: The if part: Let ζ_i be a primitive p_i-th root of unity. Then, by our assumption and by Theorem 3.10.4, each ζ_i is constructed. We may assume that the argument of ζ_i is equal to $2\pi/p_i$. Then the argument of $\zeta_1^{m_1} \cdots \zeta_t^{m_t}$ is equal to $2\pi(m_1/p_1 + \cdots + m_t/p_t)$. Since $p_1, \ldots, p_t$ are mutually distinct prime numbers, we see the existence of rational integers $m_1, \ldots, m_t$ such that the argument coincides with $2\pi/p_1 \cdots p_t$. This proves that a regular $p_1 \cdots p_t$-gon is obtained by a geometric construction. Since the bisection of an angle can be constructed, we prove the assertion for every s.

The only if part: If a regular n-gon is constructed, then for every divisor n' (≥ 3) of n, a regular n'-gon is also constructed obviously. Hence for every odd prime factor p, $p - 1$ must be a power of 2. Since $\phi(p^2) = p^2 - p$, a regular p^2-gon cannot be constructed by virtue of Corollary 3.10.3 and we see that p^2 cannot divide n. QED

COROLLARY 3.10.6 If p is an odd prime number, then a p-section of a general angle (i.e., $1/p$ of a given general angle) cannot be obtained by any geometric construction.

PROOF: A regular p^2-gon is not constructed, nor a p-section of $2\pi/p$. QED

COROLLARY 3.10.7 If the n-section of every angle is obtained by a geometric construction, then n is a power of 2, and conversely.

3.11 ALEBRAICALLY CLOSED FIELDS

A field K is said to be algebraically closed if K has no proper algebraic extension field, or equivalently , if K is a splitting field for every polynomial in one variable over K. It is well known that the complex number field C is algebraically closed; we give a proof of this later (Theorem 3.11.4). For a given field K, a field K^* is called its algebraic closure if (1) K^* is algebraic over K and (2) K^* is algebraically closed. In this case, the separable closure of K in K^* is called the separable algebraic closure of K.

As we noted after the definition of algebraic dependence, we have:

THEOREM 3.11.1 The algebraic closure K' of a field K in its extension field L is an algebraic extension of K and is algebraically closed in L. In particular, if L is algebraically closed, then K' is an algebraic closure of K.

EXAMPLE Since the complex number field C is algebraically closed, the set Q^* of complex numbers which are algebraic over Q is an algebraic closure of Q. Elements of Q^* are called algebraic numbers; complex numbers which are integral over the ring of rational integers are called algebraic integers.

THEOREM 3.11.2 Let K be an arbitrary field. (1) There is an algebraic closure K^* of K. (2) If K' is another algebraic closure of K, then K' is K-isomorphic to K^*, and hence we may identify algebraic closures of K with each other. (3) An algebraic extension L of K is an algebraic closure of K if the following condition holds good: "Every irreducible monic polynomial in X (of degree ≥ 1) over K has at least one root in L."

PROOF: (3): Letting a be an element of an algebraic extension field of L, we want to show that $a \in L$. a is algebraic over K. Let $f(X)$ be the minimal polynomial for a over K, and let S be the minimal splitting field for $f(X)$. Since S is finite and normal over K, $S = T \vee U$ with a Galois extension T and a purely inseparable extension U over K (Corollary 3.6.2). Let b, $c_1, \ldots, c_m$ be such that $T = K(b)$ (Theorem 3.4.2) and $U = K(c_1,\ldots,c_m)$. The minimal polynomial $g(X)$ for b over K has a root b' in L by our assumption. Since $K(b)$ is a Galois extension, we have $b' \in K(b)$. Since $[K(b') : K] = \deg g(X) = [K(b) : K]$, we see that $K(b) = K(b')$, which shows that $T \subseteq L$. The minimal polynomial $h_i(X)$ for c_i over K has a root in L, and the root must be c_i; hence $U \subseteq L$. Thus $a \in S \subseteq L$.

(1): We consider the <u>polynomial ring</u> P in an infinite number of variables $\{X_g \mid g \in G\}$ over K which is defined to be the union of the polynomial rings in finite numbers of X_g. For our purpose, let G be the set of irreducible monic polynomials (of degree ≥ 1) over K in a variable Y, and consider a new variable X_g for each $g = g(Y) \in G$. Let I be the ideal of P generated by $\{g(X_g) \mid g \in G\}$. If $1 \in I$, then $1 = \sum_{i=1}^{n} f_i g_i(X_{g_i})$ with $f_i \in P$, $g_i \in G$. Since this is a relation as polynomials, we can replace X_{g_i} with a root b_i of $g_i(Y)$ in a splitting field for $g_1(Y)\cdots g_n(Y)$. Then we obtain $1 = 0$, a contradiction. Therefore $1 \notin I$, and there is a maximal ideal M of P containing I. Let K^* be the field P/M. K^* is generated by the classes b_g of X_g which is a root of $g(Y)$. Therefore K^* is algebraic over K. On the other hand, every irreducible monic polynomial $g(Y)$ over K has a root b_g in K^*. Thus K^* is an algebraic closure of K by (3) proved above.

(2): Consider the set S of pairs (L,σ) of field L such that $K \subseteq L \subseteq K^*$ and a K-injection σ of the field L into K'. We introduce an order in S by $(L,\sigma) \geq (L',\sigma')$ if and only if (i) $L \supseteq L'$

and (ii) $\sigma' = \sigma|_{L'}$ (the restriction of σ on L'). We want to show that S is an inductive set. Let $\{(L_\lambda,\sigma_\lambda) \mid \lambda \in \Lambda\}$ be a well-ordered subset of S. For each $a \in L = \cup L_\lambda$, we set $a^\sigma = a^{\sigma_\lambda}$ (with λ such that $a \in L_\lambda$); then σ is a well-defined mapping and $(L,\sigma) \in S$ as is easily seen. Then we see that (L,σ) is the supremum of the well-ordered subset. Thus S is inductive, and hence by Zorn lemma, S has a maximal element, say (L^*,σ^*). Assume for a moment that $L^* \neq K^*$, and let b be an element of K^* which is not in L^*. Let f(X) be the minimal polynomial for b over L^*. Then $f^{\sigma^*}(X)$ is irreducible over $(L^*)^{\sigma^*}$, and therefore σ^* is extended to an isomorphism of $L^*(b)$ onto $(L^*)^{\sigma^*}(b')$ where b' is a root of $f^{\sigma^*}(X)$ in K'. This contradicts the maximality of (L^*,σ^*), and we see that $L^* = K^*$. QED

From the proof of (2) above, we have:

COROLLARY 3.11.3 Let K^* be the algebraic closure of a field K, and let $L \subseteq L'$ be algebraic extensions of K. Then each injection of L into K^* is extended to an injection of L' into K^*.

We add here two results on the complex number field.

THEOREM 3.11.4 The complex number field $\mathbb{C}$ is algebraically closed.

PROOF: Although the reader can find in the literature a quite natural proof based on properties of continuous functions and complex numbers, we give a proof which is based on Sylow subgroups and on the fact that every polynomial of odd degree (≥ 3) over the real number field $\mathbb{R}$ is reducible. Assume that an element a is algebraic over $\mathbb{C}$, and let L be a Galois extension of finite degree over $\mathbb{R}$ such that $\mathbb{C}(a) \subseteq L$. Consider the Galois group $G = G(L/\mathbb{R})$, and let S be a 2-Sylow subgroup. Let K be the subfield of L corresponding to S. Then $S = G(L/K)$, $\#(S) = [L : K]$ (Theorem 3.6.1); hence $[K : \mathbb{R}] = \#(G)/\#(S)$, which is odd. Since polynomials of odd degrees over $\mathbb{R}$ are reducible, we have $K = \mathbb{R}$ and $\#(G) = 2^m$ with a natural number m. If $m = 1$, then $L = \mathbb{C}$. Assume that $m \geq 2$, and let H be the subgroup of G corresponding to $\mathbb{C}$. Since H is nilpotent (Theorem 1.3.4), there is a subgroup H' such that $\#(H/H') = 2$ (Exercise 1.3.4). The field C' corresponding to H' is a quadratic extension of C and hence is of the form $C(b)$ with a root b of a quadratic polynomial $X^2 + aX + a'$ $(a, a' \in C)$; hence b is a complex number, which contradicts the existence of H'. QED

LEMMA 3.11.5 Let L be an algebraic extension field of a field K. (1) If $\#(K)$ is infinite, then $\#(L) = \#(K)$. (2) If $\#(K)$ is finite, then $\#(L)$ is countable.

PROOF: We use the notation in the proof of (1) in Theorem 3.11.2, and it suffices to prove the assertion in the case $L = K^*$. $\#(K[Y]) = \#(K)$ or countable; hence $\#(K^*) \leq \#(P) = \#(K)$ or countable. QED

THEOREM 3.11.6 There is a subset S of the real number field R such that S is <u>algebraically independent</u> (i.e., every finite subset of S is algebraically independent) over the rational number field Q and such that $\#(S) = \#(R)$.

PROOF: Consider the set of algebraically independent subsets of R over Q. It is an inductive set under the containment relation. Hence, by Zorn lemma, there is a maximal algebraically independent subset S. The set F_d of polynomials of degree at most d in elements of S over Q has cardinality $\max\{\#(Q), \#(S)\}$ (Theorem 0.1.3). Therefore, if $\#(S) < \#(R)$, then $\#(Q(S)) < \#(R)$. But, R is algebraic over $Q(S)$ by the maximality of S, and hence $\#(Q(S)) = \#(R)$ by the lemma above. Hence $\#(S) = \#(R)$. QED

EXERCISE 3.1

1. Show that if K and L are finite fields such that $K \subseteq L$, then $\#(L) = \#(K)^d$ with $d = [L : K]$.
2. Compute the degrees of extensions of the following fields over the field Q of rational number:

   (1) $Q(\sqrt{2}, \sqrt{3})$ (2) $Q(\omega)$, where $\omega^3 = 1$, $\omega \neq 1$

   (3) $Q(\sqrt[3]{2})$ (4) $Q(\sqrt{3}, \sqrt[3]{3})$
3. Define the <u>degree</u> of an algebraic element a over a field K to be the degree of the minimal polynomial for a over K. Show that if a is an element of a finite algebraic extension L of K, then the degree of a over K is a divisor of $[L : K]$.

4. Give proofs of adaptions of results in Sec. 2.9 to algebraic dependence assuming that the rings are integral domains.
5. If K is a field of characteristic $p \neq 0$ and if q is a power of p, then we denote by K^q the set $\{a^q \mid a \in K\}$. Prove that K^q is a field and that the mapping $a \rightsquigarrow a^q$ gives an isomorphism of K to K^q.
6. Prove the equality $K[a] = K(a)$ in Theorem 3.1.7 using Theorem 2.6.3.

Exercise 3.2

1. Find the minimal splitting fields and their degrees over Q of the following polynomials:
   (i) $X^4 - X^2 + 4$
   (ii) $X^3 - 2$
   (iii) $X^4 + 1$
2. Let L be the minimal splitting field of a polynomial $f(X)$ of degree n over a field K. Prove that $[L : K] \leq n!$ (i.e., $n(n - 1)\cdots 2\cdot 1$).

EXERCISE 3.3

1. Examine if each of the following polynomials has a multiple root. [We assume that the coefficients are in a field K. Only in (iv) assume that K is of characteristic $p \neq 0$. (Note that the answers depend on the characteristic of K.)]
   (i) $X^5 + X$
   (ii) $X^4 + X + 1$
   (iii) $X^4 + 2X^3 + 3X^2 + 8X + 1$
   (iv) $c_0X^{p^n} + c_1X^{p^{n-1}} + \cdots + c_iX^{p^{n-i}} + \cdots + c_{n-1}X^p + c_nX + c_{n+1}$ $(c_i \in K,\ c_0 \neq 0)$
2. Give a proof of Theorem 3.3.6 based on Theorem 3.3.5 and on the fact that the multiplicative group of a finite field K is a cyclic group of order prime to the characteristic.
3. Prove that $K(t)$ is not perfect if t is a transcendental element over a field K of characteristic $p \neq 0$.

4. Let K be a subfield of a field L. Show that the set of purely inseparable element of L over K forms a subfield of L.
5. Let K be a field of characteristic $p \neq 0$. For each natural number n, we denote by $K^{p^{-n}}$ the set of p^n-th roots of elements of K in the algebraic closure of K (cf. Sec. 3.11) and by $K^{p^{-\infty}}$ the union of all $K^{p^{-n}}$. Prove the following assertions:
   (i) $K^{p^{-n}}$ and $K^{p^{-\infty}}$ are purely inseparable over K.
   (ii) $K^{p^{-\infty}}$ is the least perfect field containing K.
   (iii) Each $K^{p^{-n}}$ is isomorphic to K, but $K^{p^{-\infty}}$ need not be isomorphic to K.
6. Let a be an algebraic element over a field K of characteristic $p \neq 0$. Show that a is separable over K if and only if $K(a) = K(a^p)$.

EXERCISE 3.4

1. Give an example of a finite algebraic extension which is not a simple extension.
2. Let L be a simple extension of a field K, and let $n = [L : K]$. Prove that the number of intermediate fields (including K and L) is at most 2^{n-1}.

EXERCISE 3.5

1. Let L be a finite algebraic extension of a field K, and let L_s be the separable closure of K in L. Prove that if L is a normal extension of K, then L_s is a Galois extension of L, and then show by an example that the converse in not true.
2. Assume that subfields $L_1, \ldots, L_m$ of a field are normal extensions of a field K. Prove that $L_1 \vee \cdots \vee L_m$ is also normal over K.
3. Show by an example that even if L and L' are normal extensions of fields K and L, respectively, L' need not be a normal extension of K.

EXERCISE 3.6

1. Find the structure of the following groups (Q is the rational number field, K is a field, and t is transcendental over K).
   (i) $\mathrm{Aut}\ Q(\sqrt[3]{2})$ (ii) $\mathrm{Aut}\ Q(\sqrt{-1}, \sqrt[4]{2})$
   (iii) $\mathrm{Aut}_{K(t^2)} K(t)$ (iv) $\mathrm{Aut}_{K(t^3)} K(t)$
2. Find the field of invariants in each of the cases above.
3. Let L be a normal extension of finite degree over a field K, and let M be an intermediate field. Set $s = [M : K]_s$ and $t = [M : K]_i$. Let $\sigma_1, \ldots, \sigma_s$ be a set of representatives for $G(L/M)\backslash G(L/K)$. Then for $a \in M$, t-times repetition of $a^{\sigma_1}, \ldots, a^{\sigma_s}$ is called the <u>complete set of conjugates</u> of a; let it be $\{a_1, \ldots, a_n\}$ $(n = st)$. Then $\sum_{i=1}^n a_i$ and $\Pi_{i=1}^n a_i$ are called the <u>trace</u> and the <u>norm</u> of a with respect to the extension M/K, and they are denoted by $Tr_{M/K}(a)$ and $N_{M/K}(a)$, respectively. Prove:
   (i) For every $a \in M$, $Tr_{M/K}(a)$ and $N_{M/K}(a)$ are in K.
   (ii) $Tr_{M/K}(a + b) = Tr_{M/K}(a) + Tr_{M/M}(b)$, $N_{M/K}(ab) = N_{M/K}(a)N_{M/K}(b)$ for arbitrary $a,b \in M$.
   (iii) Let $b_1, \ldots, b_n$ be a linearly independent base for the K-module M. For a fixed $a \in M$, consider the matrix $\rho(a)$ corresponding to the linear mapping $x \rightsquigarrow ax$ of M into M with respect to $b_1, \ldots, b_n$ [i.e., the (i, j)-entry c_{ij} of $\rho(a)$ is given by $ab_i = \sum_j c_{ij}b_j$]. Then $N_{M/K}(a) = \det \rho(a)$. Furthermore, $Tr_{M/K}(a)$ coincides with the <u>trace</u> of $\rho(a)$ (i.e. $\sum_i c_{ij}$).
   (iv) If M' is a finite algebraic extension of M, then
   $$Tr_{M/K}\cdot Tr_{M'/M} = Tr_{M'/K} \qquad N_{M/K}\cdot N_{M'/M} = N_{M'/K}$$
   (v) $Tr_{M/K}(M) = 0$ if and only if M is inseparable over K.

EXERCISE 3.7

1. Compute the number of intermediate fields of Q and $Q(\omega, \sqrt[3]{2})$ $(\omega^3 = 1, \omega \neq 1)$. Do the same for Q and $Q(\sqrt[4]{2})$.

EXERCISE 3.8

1. Prove the if part of Theorem 3.8.8.
2. If n is a power of an odd prime number, then the cyclotomic field of order n over a given field is a cyclic extension.
3. Let n be a natural number. We say that L is a Kummer n-extension if (i) K is a field such that $X^n - 1$ has n distinct root in K, (ii) L is a finite abelian extension of K, and (iii) the maximum of orders of elements of $G(L/K)$ is n. Prove the following, admitting the fact that a commutative finite group is the direct product of a finite number of cyclic groups: If L is a Kummer n-extension, then $L = K(a_1^{1/n_1},\dots,a_s^{1/n_s})$ with $a_i \in K$ and $n_1\cdots n_s = \#(G(L/K)$. Each n_i is a divisor of n.
4. Let L be a separable extension of degree n over a field K, and let $\sigma_1, \dots, \sigma_n$ be the mutually distinct K-injections of the field L into a sufficiently large field Ω. Let $b_1, \dots, b_n$ be a linearly independent base for L over K. Prove that the determinant of the $n \times n$ matrix $(b_i^{\sigma_j})$ is different from zero.
5. Assuming that K is an infinite field, generalize Exercise 4 to: By a polynomial $f(\sigma_1,\dots,\sigma_n) = \sum c_{i_1\cdots i_n}\sigma_1^{i_1}\cdots\sigma_n^{i_n}$ $(c_{i_1\cdots i_n} \in \Omega)$, we define the mapping of L into Ω such that $a \rightsquigarrow \sum c_{i_1\cdots i_n}(a^{\sigma_1})^{i_1}\cdots(a^{\sigma_n})^{i_n}$. If a polynomial $f(X_1,\dots,X_n)$ is not zero, then the mapping $f(\sigma_1,\dots,\sigma_n)$ defined as above is not zero. [Hint: Let $b_1, \dots, b_n$ be a linearly independent base for L over K. If the mapping is zero, then $f(\sum X_i b_i^{\sigma_1}, \dots, \sum X_i b_i^{\sigma_n})$ is zero as a polynomial in $X_1, \dots, X_n$. Then apply Exercise 4.]
6. Let L be an algebraic extension of finite degree over a field K. If a linearly independent base $\{b_1,\dots,b_n\}$ of L over K consists of all conjugates of b_1, then we say that this base is a normal base of L over K.

(i) Prove that if L has a normal base, then L is a Galois extension.

(ii) Assume that K is an infinite field and that L is a Galois extension of K. Let $G(L/K)$ be $\{\sigma_1, \ldots, \sigma_n\}$ ($n = [L : K]$). Prove that conjugates of an element u of L form a normal base if and only if the determinant of the $n \times n$ matrix $(u^{\sigma_i \sigma_j})$ is different from zero, and prove also the existence of such u.

7. Let L be a cyclic extension of a field K, and let σ be a generator of $G(L/K)$. Prove the following two assertions:

(i) (Theorem 90 of Hilbert) For an element a of L, $N_{L/K}(a) = 1$ if and only if there is y in L ($y \neq 0$) such that $a = y^{1-\sigma}$ ($= y/y^{\sigma}$). In this case, $a = z^{1-\sigma}$ ($0 \neq z \in L$) if and only if $zy^{-1} \in K$.

(ii) For an element a of L, $Tr_{L/K}(a) = 0$ if and only if there is b in L such that $a = b - b^{\sigma}$. In this case, $a = c - c^{\sigma}$ ($c \in L$) if and only if $b - c \in K$.

EXERCISE 3.9

1. Prove that if $L_1, \ldots, L_n$ are abelian extensions of a field K, then the field $L_1 \vee L_2 \vee \cdots \vee L_n$ is also an abelian extension of K.

2. Adapt Excercise 1 to the solvable case.

3. Find the structure of the Galois group of each of the following polynomials with rational coefficients:

(i) $X^3 + X + 1$ (ii) $X^3 - X + 1$

(iii) $X^4 + X^2 + 1$ (iv) $X^4 - X^2 + 1$

(v) $X^n - 3$ (n being a natural number)

EXERCISE 3.10

1. Let n be a natural number. Prove that the angle of $n°$ can be obtained by a geometric construction if and only if n is a multiple of 3.

2. Show that a root $X^4 + X + 1$ gives a counterexample to the converse of Lemma 3.10.3

3. Prove the impossibility of geometric construction for $a^{1/3}$ provided that a is a natural number which is not a cubic power of any natural number.

EXERCISE 3.11

1. Prove that the degree of an irreducible polynomial in X over the real number field R is at most 2, using the facts that the complex number field C is algebraically closed and that $[C : R] = 2$.

2. Let K^* be the algebraic closure of field K, and let L be the field of invariants of the group $G(K^*/K) = \mathrm{Aut}_K K^*$. Prove that L is the least perfect field containing K.

3. Prove that there is a field K such that (i) $\#(K)$ is not finite, (ii) K is algebraic over a finite field, and (iii) K is not algebraically closed.

CHAPTER 4

TRANSCENDENTAL EXTENSIONS

4.1 TRANSCENDENCE BASE

As we defined in Theorem 3.11.6, a subset S of a ring containing a field K is said to be algebraically independent over K if every finite subset of S is algebraically independent over K (i.e., if $x_1, \ldots, x_n$ are mutually distinct elements of S, then $x_1, \ldots, x_n$ are algebraically independent over K). If a field L is generated by an algebraically independent subset S over a field K, then we say that L is the rational function field of S over K or that L is purely transcendental over K. We obtained the following fact in our proof of Theorem 3.11.6:

LEMMA 4.1.1 If K is a subfield of a field L, then there is a maximal algebraically independent subset S over K and L is algebraic over $K(S)$.

Such an S is called a transcendence base of L over K. If, furthermore, L is separable over $K(S)$, then we say that S is a separating transcendence base.

THEOREM 4.1.2 If S and S' are transcendence bases of a field L over a field K, then $\#(S) = \#(S')$.

This $\#(S)$ is called the transcendence degree or the dimension of L over K and is denoted by $\text{trans.deg}_K L$. If I is a subring of an integral domain A and if K, L are fields of fractions of I, A, respectively, then $\text{trans.deg}_K L$ is called the transcendence degree of A over I.

PROOF: We assume first that $\#(S)$ is finite. Then we may assume that $n = \#(S) \le \#(S')$. Let $S = \{x_1, \ldots, x_n\}$ and let $y_1, \ldots, y_n$ be mutually distinct elements of S'. It suffices to show that if we renumber $x_1, \ldots, x_n$ suitably, then L is algebraic over $K_i = K(y_1,\ldots,y_i,x_{i+1},\ldots,x_n)$ for $i = 0, 1, \ldots, n$ (because the case $i = n$ implies that $S' = \{y_1,\ldots,y_n\}$). The case $i = 0$ is obvious, and we assume that $i \ge 1$ and that L is algebraic over K_{i-1}. Then y_i is algebraic over K_{i-1}, and we have a relation $c_0 y_i^{t-1} + \cdots + c_t = 0$ $(c_0, \ldots, c_t \in K[y_1,\ldots,y_{i-1},x_i,\ldots,x_n]$, $c_0 \neq 0)$. Since $y_1, \ldots, y_i$ are algebraically independent, some of $x_i, \ldots, x_n$ must appear in the relation. Renumbering $x_i, \ldots, x_n$ suitably, we may assume that x_i appears properly. Then x_i is algebraic over K_i, whence L is algebraic over K_i. This completes the proof of the case where $\#(S)$ is finite. Now we assume that $\#(S)$ is infinite and that $\#(S) \ge \#(S')$. Each element y of S' is algebraically dependent on a finite subset T_y of S. Set $T = \cup_y T_y$. Since every element of S' is algebraic over $K(T)$, we have $T = S$. On the other hand, $\#(S) = \#(\cup T_y) \le$ (countably infinite) $\times$ $\#(S') = \#(S')$ (Theorem 0.1.3). Thus $\#(S) = \#(S')$. QED

We note here that L is algebraic over K if and only if $\text{trans.deg}_K L = 0$.

THEOREM 4.1.3 If a field L is generated by a subset S over a field K, then there is a subset T of S which is a transcendence base of L over K.

PROOF: There is a maximal member T in the set of algebraically independent subsets of S. QED

COROLLARY 4.1.4 If a field L is finitely generated (i.e., generated by a finite number of elements) over its subfield K, then a transcendence base T of L over K is a finite set and L is of finite degree over $K(T)$.

THEOREM 4.1.5 If a field L is finitely generated over its subfield K and if M is an intermediate field, then M is finitely generated over K.

PROOF: First we assume that M is algebraic over K. Let $x_1, \ldots, x_n$ be a transcendence base of L over K. If elements $m_1, \ldots, m_r$ ($\in M$) are linearly independent over K, then they are linearly independent over $K(x_1, \ldots, x_n)$ [because a linear relation on $m_1, \ldots, m_r$ over $K(x_1, \ldots, x_n)$ is regarded as an algebraic relation on $x_1, \ldots, x_n$ over $K(m_1, \ldots, m_r)$]. Therefore $[M : K] \leq [L : K(x_1, \ldots, x_n)]$ (finite). Now we consider the general case. Let T be a transcendence base for M over K. Then by the algebraic case, M is finitely generated over $K(T)$, and hence over K, too. QED

LEMMA 4.1.6 Let K be a subfield of an algebraically closed field Ω. If L, L' are intermediate fields, if there is a K-isomorphism σ of L onto L', and if $\text{trans.deg}_K L$ is finite, then σ is extended to an automorphism of Ω.

PROOF: Let $x_1, \ldots, x_t$ be a transcendence base of L over K, and let B be a transcendence base of Ω over L. Then $A = B \cup \{x_1, \ldots, x_t\}$ is a transcendence base of Ω over K. Let $y_i = x_i^\sigma$, and let S_0 be a maximal member in the set of subsets S of A such that $\{y_1, \ldots, y_t\} \cup S$ is algebraically independent over K ($y_i \notin S$). By Theorem 4.1.3, we see that $\{y_1, \ldots, y_t\} \cup S_0$ is a transcendence base of Ω over K, and hence $\{y_1, \ldots, y_t\}$ is a transcendence base of Ω over $K(S_0)$. Therefore by virtue of Theorem 4.1.2 we have $S_0 \cup \{z_1, \ldots, z_t\} = A$ with $z_1, \ldots, z_t \in A$. Taking a one-one correspondence between B and S_0, we extended σ to a K-isomorphism of $L(B)$ onto $L'(S_0)$. Since Ω is the algebraic closure of $L(B)$ and also of $L'(S_0)$, σ is extended to an automorphism of Ω by virtue of the uniqueness of the algebraic closure (Theorem 3.11.2). QED

4.2 TENSOR PRODUCTS

Tensor products are defined for modules over a ring. But for simplicity we observe those over fields only. Let M, N be modules over a field K. Let $B = \{b_\lambda \mid \lambda \in \Lambda\}$, $C = \{c_\mu \mid \mu \in \Lambda'\}$ be linearly independent bases for M, N, respectively. The <u>tensor product</u> of

M, N over K is defined to be the K-module with the product set $B \times C$ as a linearly independent base and is denoted by $M \otimes N$, or more precisely by $M \otimes_K N$. Each (b,c) in $B \times C$ is denoted by $b \otimes c$ if it is considered as an element of $M \otimes N$. More generally, for $m = \sum a_\lambda b_\lambda$, $n = \sum d_\mu c_\mu$ (finite sums; $a_\lambda, d_\mu \in K$), we define their tensor product $m \otimes n$ to be $\sum (a_\lambda d_\mu)(b_\lambda \otimes c_\mu)$. Thus we define the canonical mapping of $M \times N$ into $M \otimes N$ $[(m,n) \rightsquigarrow m \otimes n]$.

Although we defined $M \otimes N$ depending on bases B, C, we can show uniqueness of $M \otimes N$. For our purpose, we define the notion of a bilinear mapping of the product set $M \times N$ into a K-module M^* to be a mapping f satisfying:

$$\begin{cases} f(m + m', n) = f(m,n) + f(m',n) \\ f(m, n + n') = f(m,n) + f(m,n') \\ f(am,n) = f(m,an) = a\, f(m,n) \qquad (m,m' \in M;\ n,n' \in N;\ a \in K) \end{cases}$$

THEOREM 4.2.1 (1) The canonical mapping ψ is bilinear; (2) if f is a bilinear mapping of $M \times N$ into a K-module M^*, then there is a unique K-homomorphism g such that $f = g\psi$; and (3) $M \otimes N$ is generated by $\psi(M \times N)$. Conversely, if ψ' is a bilinear mapping of $M \times N$ into a K-module L and if the pair (L,ψ') enjoys the properties of $(M \otimes N, \psi)$ in (1) to (3), then L is K-isomorphic to $M \otimes N$. In particular, the K-module structure of $M \otimes N$ is independent of the choice of the bases.

PROOF: (1) is obvious. As for (2), g such that $g(\sum a_{ij}(m_i \otimes n_j)) = \sum a_{ij} f(m_i,n_j)$ is a homomorphism such that $f = g\psi$. Uniqueness of g is clear. (3) is obvious. Consider (L,ψ') as in the last assertion. Since ψ is bilinear, there is a K-homomorphism g^* of L into $M \otimes N$ such that $\psi = g^*\psi'$. g^* is surjective by property (3). Similarly, there is a K-homomorphism g of $M \otimes N$ onto L such that $\psi' = g\psi$. Consider g^*g, which is a K-surjection of $M \otimes N$ to itself, and $\psi = g^*g\psi$. Therefore, by property (2), we see that g^*g is the identity mapping. Similarly, gg^* is the identity mapping of L. Thus we see that g is an isomorphism of $M \otimes N$ to L. QED

THEOREM 4.2.2 Assume that M, N, L are modules over a field K.

(1) $M \otimes_K N \cong N \otimes_K M$, and $(M \otimes_K N) \otimes_K L \cong M \otimes_K (N \otimes_K L)$.

(2) If N is a ring, then $M \otimes N$ is an N-module by defining $\sum(m_i \otimes n_i)n = \sum m_i \otimes (n_i n)$.

(3) If M, N are rings, then $M \otimes N$ is a ring by defining $(\sum m_i \otimes n_i)(\sum p_j \otimes q_j) = \sum (m_i p_j) \otimes (n_i q_j)$.

(4) If M', N' are submodules of M, N, respectively, then $M' \otimes N'$ is naturally a submodule of $M \otimes N$.†

PROOF: (1) to (3) are proved easily. We see (4) observing bases of M, N which contain bases of M', N', respectively. QED

If M, N are rings having common identity with a subfield K, then we define canonical mappings ψ, ϕ of M, N into $M \otimes_K N$ by $\psi m = m \otimes 1$, $\phi n = 1 \otimes n$. Then they are injections as is easily seen. Therefore we often identity M, N with ψM, ϕN, respectively.

THEOREM 4.2.3 If furthermore M, N are subrings of a ring, say R, then there is a homomorphism f of the ring $M \otimes_K N$ onto the ring M[N] generated by M and N such that $f(m \otimes n) = mn$.

This f is called the natural homomorphism of $M \otimes N$ into R.

The proof is easy and we leave it to the readers.

We add here two more definitions. If M, N are fields containing a field K, then the total quotient ring of $M \otimes_K N$ is called the local tensor product of M, N over K and is denoted by $M \times_K N$. If furthermore M, N are subfields of a field and if the natural homomorphism f of $M \otimes N$ into the field $M \vee N$ is an injection, then we say that M, N are linearly disjoint over K. In this case f extends to an isomorphism of $M \times_K N$ to $M \vee N$ which is called the natural mapping of $M \times_K N$ to $M \vee N$.

THEOREM 4.2.4 Let M be a module over a field K and let N be a ring containing K and having the same identity with K. M is identified with $M \otimes 1$ in $M \otimes N$ via canonical mapping. Then:

---

† This is a speciality of the case over a field.

(1) If n $(\in N)$ is a nonzero-divisor in N, then n is not a zero-divisor with respect to the N-module $M \otimes N$.

(2) A linearly independent base of M over K is a linearly independent base of $M \otimes N$ over N, and hence $M \otimes N$ is a free N-module.

In particular, if N is a field, then $\text{length}_K M = \text{length}_N M \otimes N$.

(3) If N is a field and if L is an N-module (hence is a K-module naturally), then $M \otimes_K L \cong (M \otimes_K N) \otimes_N L$.

PROOF: (1): Set $N' = N[n^{-1}]$. Then $M \otimes N \subseteq M \otimes N'$ (Theorem 4.2.2). Hence $g \in M \otimes N$, $gn = 0$ imply that $g = (gn)n^{-1} = 0$. (2) is obvious. (3): Let $\{b_\lambda \mid \lambda \in \Lambda\}$, $\{c_\mu \mid \mu \in \Lambda'\}$ be linearly independent bases of N, L over K, N, respectively. Then $\{b_\lambda c_\mu \mid \lambda \in \Lambda, \mu \in \Lambda'\}$ forms a linearly independent base of L over K. Therefore we see the existence of an isomorphism such that $m \otimes b_\lambda \otimes c_\mu \rightsquigarrow m \otimes (b_\lambda c_\mu)$. QED

LEMMA 4.2.5 If a ring R is a finitely generated module over its subfield K, then nonzero-divisors of R are units in R.

PROOF: Let a be a nonzero-divisor. Since $\text{length}_K R$ is finite, there is a natural number d such that $1, a, \ldots, a^{d-1}$ are linearly independent and a^d is linearly dependent on them. Then $a^d = c_0 + c_1 a + \cdots + c_{d-1}a^{d-1}$ with $c_i \in K$. If $c_0 = 0$, then $a(a^{d-1} - c_1 - \cdots - c_{d-1}a^{d-2}) = 0$. Since a is not a zero-divisor, we have $a^{d-1} = c_1 + \cdots + c_{d-1}a^{d-2}$, a contradiction. Thus $c_0 \neq 0$. Then, in the total quotient ring of R, we have a relation $a^{d-1} = c_0 a^{-1} + c_1 + \cdots + c_{d-1}a^{d-2}$. Since K is a field and $c_0 \neq 0$, we see that $a^{-1} \in K[a] \subseteq R$. QED

THEOREM 4.2.6 If M, N are extension fields of a field K and if N is algebraic over K, then $M \times_K N = M \otimes_K N$.

PROOF: It suffices to show that each nonzero-divisor $f = \sum_{i=1}^{s} m_i \otimes n_i$ is a unit in $M \otimes N$. Let $N' = K(n_1,\ldots,n_s)$. Then $f \in M \otimes N' \subseteq M \otimes N$. On the other hand, $M \otimes N'$ is finite over the field M; hence f is a unit in $M \otimes N'$ by the lemma above. QED

THEOREM 4.2.7 Let M, N be extension fields of a field K. Let T be a transcendence base of N over K, and set $L = K(T)$. Then (1) $M \times_K L$ is a subfield of $M \times_K N$; and (2) if $[N : L] = d$ (finite), then $M \times_K N$ has length d as $(M \times_K L)$-module, and hence is an Artin ring.

PROOF: (1): Let P be the ring generated by T over K. Then P is isomorphic to the polynomial ring over K, and monomials in elements of T form a free base. Therefore we see that $M \otimes P$ is regarded as the polynomial ring in elements of T over M. Thus $M \otimes P$ is an integral domain and $M \times_K L$ is a field. Furthermore, Theorem 4.2.4 applied to $(M \otimes_K L) \otimes_L N$ shows that nonzero-divisors in $M \otimes L$ (this is an integral domain because $M \otimes L \subseteq M \times_K L$) are nonzero-divisors in $M \otimes N$. Thus $M \times_K L \subseteq M \times_K N$. (2): Since $(M \times_K L) \otimes_L N = (M \times_K L) \times_L N$ by the last theorem, we see that $(M \times_K L) \otimes_L N = M \times_K N$, and we see (2). QED

THEOREM 4.2.8 Let M, N be extension fields of a field K which generate a field $M \vee N$, and let P be the kernel of the natural homomorphism ψ of $M \otimes_K N$ into $M \vee N$.

(1) Every prime ideal of $M \times_K N$ is maximal.

(2) The following (i) and (ii) are equivalent to each other:

(i) There is a maximal ideal Q of $M \times_K N$ such that $P = Q \cap (M \otimes N)$.

(ii) For every finitely generated subfield M' of M over K, it holds that $\text{trans.deg}_K M' = \text{trans.deg}_N N \vee M'$.

If condition (i) or (ii) is satisfied, then we say that M, N are <u>free</u> over K.

PROOF: Assume that there is a prime ideal Q of $M \times_K N$ which is not maximal. Let Q^* be a maximal ideal containing Q, and let a be an element of Q^* which is not in Q. Then there is a finitely generated subfield M' of M such that $a \in M' \times_K N$. $Q \cap (M' \times_K N)$ and $Q^* \cap (M' \times_K N)$ are prime ideals of $M' \times_K N$; this contradicts the fact that $M' \times_K N$ is an Artin ring (Theorem 4.2.7). Thus (1) is proved. As for (2), it is obvious that (i) implies (ii). Assume that (i) does not hold. It means that there are finitely generated subfields M', N' of M, N such that the kernel of the restriction of ψ on $M' \otimes N'$ generates the unit ideal in $M' \times_K N'$. Thus we may assume that M is finitely generated. Let $x_1, \ldots, x_t$ be a transcendence base of M over K. The set S of nonzero elements of $K[x_1,\ldots,x_t] \otimes N$ consists only of non-zero-divisors (Theorem 4.2.4). Hence we can consider the subring T of $M \times_K N$ generated by inverses of elements of S over $M \otimes N$ (cf. Theorem 4.6.12). Then T contains $N(x_1,\ldots,x_t)$ $(\subseteq M \times_K N)$, and T is a finite $N(x_1,\ldots,x_t)$-module. Therefore T is an Artin ring, and $T = M \times_K N$. Our assumption that $1 \in P(M \times_K N)$ means that there are $p_1, \ldots, p_s$ $\in P$ and $b_1, \ldots, b_s \in M \times_K N$ such that $\sum p_i b_i = 1$. Since $T =$ $M \times_K N$, we see that there is y in S such that $yb_i \in M \otimes N$, namely that $P \cap S$ contains y $(= \sum y b_i p_i)$. Therefore $x_1, \ldots, x_t$ are algebraically dependent over N, whence (ii) does not hold. QED

4.3 DERIVATIONS

A <u>derivation</u> of a field K is a mapping D of K into K such that (1) $D(f + g) = Df + Dg$ and (2) $D(fg) = f(Dg) + g(Df)$ (for arbitrary $f,g \in K$). If, in this case, S is a subset of K such that $DS = 0$ (i.e., $Ds = 0$ for every $s \in S$), then we say that D is a derivation <u>over</u> S. For a given subset S of a field K, the set of derivations of K over S forms a K-module under operations defined by $(D + D')f = Df + D'f$, $(fD)g = f(Dg)$. We shall denote this K-module by $\operatorname{Der}(K/S)$.

THEOREM 4.3.1 If D is a derivation of a field K, then (i) $D(-f) = -Df$ for every $f \in K$; (ii) $S = \{f \in K \mid Df = 0\}$ forms a subfield: (iii) $D(f/g) = (gDf - fDg)/g^2$ $(f,g \in K,\ g \neq 0)$; (iv) $Df^n = nf^{n-1}Df$ (for $0 \neq f \in K$ and rational integer n); and (v) if the characteristic p of K is different from zero, then S in (ii) contains $K^p = \{f^p \mid f \in K\}$.

PROOF: Since $f = (f/g)g$, we have $Df = gD(f/g) + (f/g)Dg$ and we have (iii). Since $1^2 = 1$, we have $D1 = D1 + D1$, and hence $D1 =$ 0. Thus $1 \in S$. Since $0 + 0 = 0$, we have $D0 = 0$ and $0 \in S$. Since $f + (-f) = 0$, we have $Df + D(-f) = 0$ and $D(-f) = -Df$. Thus we have

(i) and particularly that $f \in S$ implies $-f \in S$. If $f,g \in S$, $g \neq 0$, then (iii) implies that $D(f/g) = 0$ and $f/g \in S$. Thus S is a field. As for (iv), if n is 0 or 1, then the assertion is obvious. If $n \geq 2$, then $Df^n = D(f \cdot f^{n-1}) = f^{n-1}Df + fDf^{n-1}$, and we prove the required equality by induction on n. If $n < 0$, then, with $m = -n$, we have $Df^n = D(1/f^m) = (f^m D1 - Df^m)/f^{2m} = (-Df^m)/f^{2m} = -mf^{m-1}(Df)/f^{2m} = nf^{n-1}Df$. (v) follows from the fact that $Df^p = p(f^{p-1}Df) = 0$. QED

THEOREM 4.3.2 Let K be the field of fractions of an integral domain I. A mapping d of I into K is the restriction of some derivation D of K if and only if d enjoys the properties (1) $d(f + g) = df + dg$ and (2) $d(fg) = fdg + gdf$ (for arbitrary f,g I). In such a case D is uniquely determined by d.

PROOF: Necessity is obvious. Sufficiency: If $f/g = f'/g'$, then $fg' = f'g$ and $fdg' + g'df = f'dg + gdf'$. Hence $g'df - f'dg = gdf' - fdg'$ and $(df - (f/g)dg)/g = (df - (f'/g')dg)/g = (df' - (f/g)dg')/g' = (df' - (f'/g')dg')/g'$. This shows that the mapping D such that $D(f/g) = (gdf - fdg)/g^2$ is well defined. One sees easily that this D is a derivation of K and d is the restriction of D. Uniqueness of D follows from (iii) in the last theorem. QED

In Chap. 3, we defined derivations of polynomials. This is generalized to polynomials in many variables. Namely, we consider the polynomial ring $K[X_1,\ldots,X_n]$ over an integral domain K. Then we can consider the mapping D_i of K which maps each polynomial $f(X_1,\ldots,X_n)$ to its derivative as a polynomial in X_i over $K[X_1, \ldots,X_{i-1},X_{i+1},\ldots,X_n]$. Then these $D_1, \ldots, D_n$ enjoy the properties in Theorem 4.3.2; hence they define derivations of $K(X_1,\ldots,X_n)$. These D_i are denoted by the symbols $\frac{\partial}{\partial X_i}$ (or $\partial/\partial X_i$) and are called <u>partial derivations</u>. $(\partial/\partial X_i)f$ is denoted by $\partial f/\partial X_i$.

On the other hand, assume that K is a field and that D is a derivation of K. Then D defines a mapping D' of the polynomial ring so that $D'(\sum c_{i_1\cdots i_n} X_1^{i_1}\cdots X_n^{i_n}) = \sum (Dc_{i_1\cdots i_n})X_1^{i_1}\cdots X_n^{i_n}$.

This D' enjoys the properties for d in Theorem 4.3.2; hence D defines a derivation D^* of $K(X_1, \ldots, X_n)$ and D, D' are restrictions of D^*. These D' and D^* are called the trivial extensions of D to $K[X_1,\ldots,X_n]$ and $K(X_1,\ldots,X_n)$, respectively, and they are denoted by the same symbol D.

THEOREM 4.3.3 Assume that a field L is generated by elements $a_1, \ldots, a_n$ over a field K. Let D be a derivation of K, and let $v_1, \ldots, v_n$ be elements of L. We consider the polynomial ring $K[X_1,\ldots,X_n]$ and the relation ideal A for $a_1, \ldots, a_n$. Let $F = \{f_1,\ldots,f_s,\ldots\}$ be a set of generators for A. For each g in $K[X_1, \ldots,X_n]$, we define H_g to be the polynomial $Dg + \sum_{i=1}^n (\partial g/\partial X_i)v_i$. Then there is an extension D' of D to a derivation of L such that $D'a_i = v_i$ for every i if and only if $H_f(a_1,\ldots,a_n) = 0$ for every $f \in F$.

PROOF: If such D' exists, then since $f(a_1,\ldots,a_n) = 0$ for $f \in F$, we have $H_f(a_1,\ldots,a_n) = D'(f(a_1,\ldots,a_n)) = 0$. Conversely, assume the validity of $H_f(a_1,\ldots,a_n) = 0$. If $f \in A$, then $f = \sum_{i=1}^s h_i f_i$ with $f_i \in F$. Then $H_f = \sum h_i Df_i + \sum f_i Dh_i + \sum_{i,j} h_i(\partial f_i/\partial X_j)v_j + \sum_{i,j} f_i(\partial h_i/\partial X_j)v_j$, and we have $H_f(a_1,\ldots,a_n) =0$. Therefore, if $g,h \in K[X_1,\ldots,X_n]$ and if $g(a_1,\ldots,a_n) = h(a_1, \ldots,a_n)$, then $g - h \in A$ and therefore $H_g(a_1,\ldots,a_n) = H_h(a_1,\ldots,a_n)$. D' such that $D'(g(a_1,\ldots a_n)) = H_g(a_1,\ldots,a_n)$ gives a well-defined mapping of $K[a_1,\ldots,a_n]$ to itself. This D' enjoys the properties for d in Theorem 4.3.2, and we see the existence of an extension of D. QED

In general, when $f_1, \ldots, f_m$ are in $K(X_1,\ldots,X_n)$, then the $m \times n$ matrix $(\partial f_i/\partial X_j)$ is called the jacobian matrix of $f_1, \ldots, f_m$ and is denoted by $J(f_1,\ldots,f_m)$. This may be adapted to infinitely many f_i. Then, under the notation in Theorem 4.3.3, we have the following important result.

THEOREM 4.3.4 Extensions D' of D to derivations of L have one-one correspondence to solutions of the system of linear equation:

$$-B\begin{pmatrix} v_1 \\ \cdot \\ \cdot \\ \cdot \\ v_n \end{pmatrix} = \begin{pmatrix} (Df_1)(a_1,\ldots,a_n) \\ \cdots\cdots\cdots\cdots \\ (Df_m)(a_1,\ldots,a_n) \\ \cdots\cdots\cdots\cdots \end{pmatrix}$$

where $B = J(f_1,\ldots,f_m,\ldots)_{(X_i=a_i)}$ which means the matrix obtained from $J(f_1,\ldots,f_m,\ldots)$ by setting $X_i = a_i$ for $i = 1, \ldots, n$.

In particular, (1) if rank $B = n$, then D' exists and is unique, and (2) the L-module $\mathrm{Der}(L/K)$ has length $(n - \text{rank } B)$.

PROOF: The first part follows from the proof of Theorem 4.3.3. (1) is immediate. As for (2), elements of $\mathrm{Der}(L/K)$ are extensions of trivial derivations of K, and hence they correspond to solutions in case Df_i are all zero. QED

THEOREM 4.3.5 Let D be a derivation of a field K, and let $L = K(a)$ be a simple extension of K.

(1) If a is transcendental over K, then for each $v \in L$, there is an extension D' of D to a derivation of L such that $D'a = v$.

(2) If a is separably algebraic over K, then an extension D' exists and is unique.

(3) Assume that K is of characteristic $p \neq 0$ and that $a^p \in K$, $a \notin K$. Then there is an extension D' of D to a derivation of L if and only if $D(a^p) = 0$. If this condition is satisfied, then for every $v \in L$ there is such a D' with $D'a = v$.

PROOF: Consider the relation ideal A for a in the polynomial ring $K[X]$. Then we can apply Theorems 4.3.3 and 4.3.4. (1) is obvious because $A = 0$. In case (2), A is generated by a monic polynomial $f(X)$ whose derivative does not vanish at $X = a$ (Lemma 3.3.1). Therefore (2) follows from Theorem 4.3.4, (1). In the case (3), A is generated by $f(X) = X^p - a^p$. First of all, if D' exists, then $D'a^p = 0$ and the only if part is obvious. Assume that $Da^p = 0$, and let $v \in L$. Then $Df = 0$ and $\partial f/\partial X = 0$. Therefore we see the existence of D' by Theorem 4.3.3. QED

THEOREM 4.3.6 If $X_1, \ldots, X_n$ are algebraically independent elements over a field K, then the partial derivations $\partial/\partial X_i$ ($i = 1, \ldots, n$) form a linearly independent base of the $K(X_1,\ldots,X_n)$-module $\mathrm{Der}(K(X_1,\ldots,X_n)/K)$.

PROOF: Let $D \in \mathrm{Der}(K(X_1,\ldots,X_n)/K)$ and $DX_i = u_i$ ($i = 1, \ldots, n$). Set $D' = D - \sum u_i(\partial/\partial X_i)$. Then $D'X_i = 0$ and D' is trivial on $K(X_1,\ldots,X_n)$. Thus $D = \sum u_i(\partial/\partial X_i)$. If $D = \sum v_i(\partial/\partial X_i)$, then $v_i = DX_i = u_i$, and therefore $\partial/\partial X_1, \ldots, \partial/\partial X_n$ are linearly independent. QED

We say that a subset S of a field K of characteristic p is <u>p-independent</u> over its subfield k if for any finite number of mutually distinct elements $s_1, \ldots, s_m$ of S it holds that $[L(s_1,\ldots,s_m) : L] = p^m$ where $L = K^p \vee k$. A maximal p-independent subset of K over k is called a p-<u>base</u> of K over k.

THEOREM 4.3.7 Under the circumstances, for any given p-independent subset S of K over k, there is a p-base S^* of K over k such that $S \subseteq S^*$. Let H be the set of mappings of S^* into K. Then there is a one-one correspondence between H and $\mathrm{Der}(K/k)$ such that ψ corresponds to D if and only if $Ds = \psi s$ for all $s \in S^*$. In particular, if $\#(S^*) = m < \infty$, then $\mathrm{length}_K \mathrm{Der}(K/k) = m$.

PROOF: The existence of S^* is easy by Zorn lemma. We now prove the last assertion. Set $L_i = L(s_1,\ldots,s_{i-1},s_{i+1},\ldots,s_m)$ ($i = 1, 2, \ldots, m$). Since $[K : L_i] = p$, there is a derivation D_i of K over L_i such that $D_i s_i = 1$ (Theorem 4.3.5). Then we see that $D_1, \ldots, D_m$ form a linearly independent base of $\mathrm{Der}(K/k)$ as in our proof of Theorem 4.3.6, and we have $\mathrm{length}_K \mathrm{Der}(K/k) = m$. With regard to the second assertion, let σ be the mapping of $\mathrm{Der}(K/k)$ into H such that $(\sigma D)s = Ds$ for all $s \in S^*$. It suffices to show that σ is a surjection. Let ψ be an arbitrary element of H. We define a mapping D of K into K as follows: For each $a \in K$, take $s_1, \ldots, s_m$ so that $a \in L(s_1,\ldots,s_m)$. Then take D' from $\mathrm{Der}(L(s_1,\ldots,s_m)/L)$ such that $D's_i = \psi s_i$. Then define Da to be $D'a$. We see easily that $D'a$ does not depend on the choice of s_1,

..., $s_m \in S$ in view of the fact that a is expressed as polynomial in elements of S with coefficients in L so that the degree in each $s \in S$ is less than p and that such an expression of a is unique. Obviously this D is an element of $Der(K/k)$. QED

4.4 SEPARABLE EXTENSIONS

Here and in the next section, the symbol p means: (1) p denotes the characteristic, if we are observing a field of nonzero characteristic. (2) $p = 1$, if we are observing a field of characteristic zero.

In case $p = 1$, $K^{p^{-n}}$ and $K^{p^{-\infty}}$ are K itself; otherwise we follow the notation in Exercise 3.3.5.

We say that a field L is <u>separable</u> or <u>separably generated</u> over its subfied K if $L \otimes_K K^{p^{-1}}$ is an integral domain. Note that in such a case $L \otimes K^{p^{-1}}$ is a field (Theorem 4.2.6). On the other hand, it is obvious that:

THEOREM 4.4.1 If a field L is separable over its subfield K, then every intermediate field is separable over K.

The following characterizations of separability include the fact that our new definition agrees with old one (in algebraic case).

THEOREM 4.4.2 If a field L is finitely generated over its subfield K, then the following five conditions are equivalent to each other:

(1) L is separable over K.

(2) L has a separating transcendence base over K.

(3) $\text{length}_L \, Der(L/K) = \text{trans.deg}_K \, L$.

(4) $\text{length}_L \, Der(L/K) \leq \text{trans.deg}_K \, L$.

(5) $L \otimes_K K^{p^{-r}}$ is an integral domain for a given natural number r.

PROOF: (2) $\Rightarrow$ (3): Assume that $x_1, \ldots, x_n$ form a separating transcendence base of L over K. Then $Der(K(x_1,\ldots,x_n)/K)$ has

$\partial/\partial x_1, \ldots, \partial/\partial x_n$ as a linearly independent base (Theorem 4.3.6). Since L is separably algebraic over $K(x_1,\ldots,x_n)$, these partial derivations are extended uniquely to derivations of L (Theorem 4.3.5) and they form a linearly independent base of Der(L/K) (cf. the proof of Theorem 4.3.6).

The implication (3) $\Rightarrow$ (4) is obvious.

(4) $\Rightarrow$ (2): (2) is obvious in the zero characteristic case, and we assume that $p \geq 2$. Let $s_1, \ldots, s_n$ be a p-base of L over K. Then each element D of Der(L/K) is determined by Ds_i ($i = 1, \ldots, n$) (Theorem 4.3.7). Therefore on one hand $\text{length}_K \text{Der}(L/K) = n$, and on the other hand $\text{Der}(L/K(s_1,\ldots,s_n)) = 0$. Since $n \leq \text{trans.deg}_K L$ by our assumption, it suffices to show that L is separably algebraic over $K^* = K(s_1,\ldots,s_n)$. Assume first that L is algebraic over K^*. Let K' be the separable closure of K^* in L. Then $\text{Der}(L/K') \subseteq \text{Der}(L/K^*)$, and hence $\text{Der}(L/K') = 0$, which implies $L = K'$ by virtue of Theorem 4.3.5. Assume next that L is transcendental over K^*. Take a transcendence base T of L over K^*. Then L is separable over $K^*(T)$ just as in the algebraic case. Each nontrivial element D of $\text{Der}(K^*(T)/K)$ is extended to a derivation of L (Theorem 4.3.5). This contradicts the fact that $\text{Der}(L/K^*) = 0$. Thus L is separably algebraic over K^*, and the implication is proved.

(2) $\Rightarrow$ (5): Take a separating transcendence base $x_1, \ldots, x_n$ of L over K. Since L is separably algebraic over $K^* = K(x_1, \ldots, x_n)$ and since $L^* = K^* \vee K^{p^{-r}}$ is normal over K^*, we see that $[L : K^*] = [L \vee L^* : L^*]$ (Theorem 3.6.3). Now, $L \otimes_K K^{p^{-r}} \cong L \otimes_{K^*} (K^* \otimes_K K^{p^{-r}}) \subseteq L \otimes_{K^*} (K^* \times_K K^{p^{-r}}) = L \otimes_{K^*} L^*$. This last ring is of length $[L : K^*]$ as an L^*-module. The natural mapping of this last ring in $L \vee K^{p^{-r}} = L \vee L^*$ is surjective. Therefore $[L : K^*] = [L \vee L^* : L^*]$ implies that the natural mapping is an isomorphism, which proves that $L \otimes K^{p^{-r}}$ is an integral domain.

The implication (5) $\Rightarrow$ (1) follows from the fact that $L \otimes K^{p^{-1}} \subseteq L \otimes K^{p^{-r}}$.

(1) ⟹ (2): As in (4) ⟹ (2), we may assume that $p \geq 2$ and we take $s_1, \ldots, s_n$ similarly. Assume for a moment that $s_1, \ldots, s_n$ are algebraically dependent over K, hence that there is a nonzero polynomial $f(X_1,\ldots,X_n) \in K[X_1,\ldots,X_n]$ such that $f(s_1,\ldots,s_n) = 0$. Choose such an f to be of smallest degree. Since $s_1, \ldots, s_n$ are p-independent over K, rearrangement of the relation $f(s_1,\ldots,s_n)$ to be a relation over $K \vee L^p$ must be trivial. This means that f is a polynomial in $X_1{}^p, \ldots, X_n{}^p$. Thus $f = g^p$ with a polynomial g over $K^{p^{-1}}$. Let g be $\sum c_{i_1\cdots i_n} X_1{}^{i_1}\cdots X_n{}^{i_n}$. Then the element $b = \sum(s_1{}^{i_1}\cdots s_n{}^{i_n}) \otimes c_{i_1\cdots i_n}$ of $L \otimes K^{p^{-1}}$ is mapped to zero under the natural homomorphism of $L \otimes K^{p^{-1}}$ to $L \vee K^{p^{-1}}$. But $L \otimes K^{p^{-1}} = K^{p^{-1}}[X_1,\ldots,X_n]/A$ with the ideal A generated by the relation ideal for $s_1, \ldots, s_n$ over K. Therefore by the minimality of the degree of f, we see that $b \neq 0$. Therefore $L \otimes K^{p^{-1}}$ is not a field, which shows that $L \otimes K^{p^{-1}}$ is not an integral domain (Theorem 4.2.7). QED

Although the notion of the order of inseparability of an extension field can be treated in a more general way, we define the notion in a simple case and show some properties of the notion.

THEOREM 4.4.3 If a field L is finitely generated over its subfield K, then $L \otimes_K K^{p^{-\infty}}$ is an Artin ring having only one maximal ideal. If $x_1, \ldots, x_n$ form a transcendence base of L over K, then the length r of this ring coincides with the ratio $[L : K(x_1, \ldots, x_n)]/[L \vee K^{p^{-\infty}} : K^{p^{-\infty}}(x_1,\ldots,x_n)]$. Therefore r is a power of p.

This r is called the <u>order of inseparability</u> of L over K and is denoted by $i(L/K)$.

PROOF: Assume that $f = \sum a_i \otimes b_i$ is a nonunit in $L \otimes K^{p^{-\infty}}$. There is a natural number t such that $b_i{}^{p^t} \in K$ for every i. Then

$f^{p^t} = \sum a_i^{p^t} \otimes b_i^{p^t} = \sum (a_i^{p^t} b_i^{p^t}) \otimes 1$, and this is nonunit. Therefore $f^{p^t} = 0$. Thus every maximal ideal contains all nonunits, and $L \otimes K^{p^{-\infty}}$ has only one maximal ideal, say M. On the other hand, $L \otimes K^{p^{-\infty}} \cong L \otimes_{K(x_1,\ldots,x_n)} (K(x_1,\ldots,x_n) \otimes_K K^{p^{-\infty}}) \cong L \otimes_{K(x_1,\ldots,x_n)} K^{p^{-\infty}}(x_1, \ldots, x_n)$. Thus $L \otimes K^{p^{-\infty}}$ is a $K^{p^{-\infty}}(x_1,\ldots,x_n)$-module, and its length coincides with $[L : K(x_1,\ldots,x_n)]$. Since $L \otimes K^{p^{-\infty}}/M \cong L \vee K^{p^{-\infty}}$, we see the coincidence of r with the ratio by virtue of Theorem 2.7.7. QED

THEOREM 4.4.4 Assume that a field L is finitely generated over its subfield K. (1) L is separable over K if and only if $i(L/K) = 1$. In this case $i(L \vee K^{p^{-n}}/K^{p^{-n}}) = 1$ $(n = 1, 2, \ldots)$. (2) If $i(L/K) > 1$, then $i(L \vee K^{p^{-1}}/K^{p^{-1}}) < i(L/K)$. (3) If $i(L/K) = p^e$, then $L \vee K^{p^{-e}}$ is separable over $K^{p^{-e}}$.

PROOF: As for (1), the first half follows from the fact that $i(L/K) = 1$ if and only if $L \otimes K^{p^{-\infty}}$ is a field; the rest follows from the fact that there is a natural surjection ψ_n of $L \otimes K^{p^{-\infty}}$ to $(L \vee K^{p^{-n}}) \otimes_{K^{p^{-n}}} K^{p^{-\infty}}$. (2) follows from the fact that the kernel of ψ_1 contains the maximal ideal of $L \otimes K^{p^{-1}}$ $(\subseteq L \otimes K^{p^{-\infty}})$. (3) follows from (2) by an induction on e. QED

COROLLARY 4.4.5 If a field L is finitely generated over a field K, then there is a purely inseparable extension K' of finite degree over K such that $L \vee K'$ is separable over K'.

PROOF: If $i(L/K) = 1$, we may set $K' = K$. Assume that $i(L/K) > 1$ and let $\sum c_i \otimes a_i$ be a nonzero nilpotent element of $L \otimes K^{p^{-1}}$. Let K_1 be the field generated by these a_i over K. Then $i(L \vee K_1/K_1) < i(L/K)$ by the proof of Theorem 4.4.4, and we complete the proof by induction on $i(L/K)$. QED

4.5 REGULAR EXTENSIONS

As we noted in Sec. 4.4, in this secion p will denote the characteristic of the field of consideration or $p = 1$ in the zero characteristic case.

A field L is said to be a <u>regular extension</u> of its subfield K if $L \otimes_K K^*$ is an integral domain, where K^* is the algebraic closure of K. This definition implies:

THEOREM 4.5.1 If a field L is a regular extension of a field K, then every intermediate field is a regular extension of K.

THEOREM 4.5.2 For an extension field L of a field K, each of the following two is equivalent to the regularity of the extension.

(1) L is separable over K and K is algebraically closed in L.

(2) $L \otimes_K L'$ is an integral domain for every extension field L' of K.

PROOF: Assume first that L is a regular extension of K. Since $K^{p^{-1}} \subseteq K^*$, L is obviously separable over K. Assume that a $(\in L)$ is algebraic over K. Let $f(X)$ be the minimal polynomial for a over K. Then $L \otimes_K K(a) = L[X]/f(X)L[X]$, which must be an integral domain because $L(a) \subseteq K^*$. Since $a \in L$, we see that $f(X) = X - a$ and $a \in K$. Thus (1) follows from the regularity.

(1) $\Rightarrow$ (2): Assume that (1) holds good but (2) does not. Then there is an extension field L' such that $L \otimes L'$ is not an integral domain. Then there are two nonzero elements $\sum a_i \otimes b_i$ and $\sum c_j \otimes d_j$ whose product is zero. Then considering the subfield generated by these b_i, d_j, we may assume that L' is finitely generated over K. Among such L', we choose one whose transcendence degree over K is the smallest; let it be M. Similarly, there is a finitely generated subfield L_1 of L over K such that $L_1 \otimes M$ is not an integral domain. Then L_1 satisfies the condition (1). Therefore L_1 has a separating transcendence base $x_1, \ldots, x_n$ (Theorem 4.4.2). Take a transcendence base $y_1, \ldots, y_m$ of M over K. Then $L_1 \otimes_K M$ is finite over $L_1(y_1,\ldots,y_m) = L_1 \otimes_K K(y_1,\ldots,y_m)$. Let P be a

maximal ideal of $L_1 \otimes_K M$ and let ψ be the natural homomorphism of $L_1 \otimes_K M$ to $L_1 \otimes_K M/P$. Take a nonzero element $f = \sum_{i=1}^{s} a_i \otimes b_i$ of P; we express it so that $a_1, \ldots, a_s$ are linearly independent. We rechoose such f so that the number s of the terms is smallest and $b_1 = 1$. We regard $\psi(L_1 \times_K M)$ as $L_1 \vee M$, and in this sense we consider $K(b_2,\ldots,b_s)$. (i) If $K(b_2,\ldots,b_s)$ is not separably algebraic over K, then there is a nontrivial D in $\mathrm{Der}(K(b_2,\ldots,b_s)/K)$. Since $x_1, \ldots, x_n$ form a separating transcendence base of L_1 over K, they form such a base of $L_1 \vee K(b_2,\ldots,b_s)$ over $K(b_2,\ldots,b_s)$. Therefore D is extended to a derivation of $L_1(b_2, \ldots,b_s)$ so that $Dx_i = 0$ for all i. Then D becomes a derivation over L_1. Since $\psi(\sum a_i \otimes b_i) = 0$, we have $\sum a_i b_i = 0$. Thus $0 = D(\sum a_i b_i) = \sum a_i Db_i$. Since $b_1 = 1$, the last expression has terms less than s; since D is nontrivial on $K(b_2,\ldots,b_s)$, this last gives a nontrivial sum. This contradicts the minimality of s among elements of P. (ii) Assume now that $K(b_2,\ldots,b_s)$ is separably algebraic over K. Since f is a nonunit in $L_1 \otimes K(b_2,\ldots,b_s) = L_1 \times_K K(b_2,\ldots,b_s)$, we may set $M = K(b_2,\ldots,b_s)$. Let M^* be a Galois extension of finite degree over K such that $M \subseteq M^*$. Since K is algebraically closed in L_1, it holds that $L_1 \cap M^* = K$. Therefore Theorem 3.6.3 implies that $[L_1 \vee M^* : L_1] = [M^* : K]$; this implies, in view of $\mathrm{length}_{L_1} L_1 \otimes M^* = [M^* : K]$, that the natural homomorphism of $L_1 \otimes M^*$ onto $L_1 \vee M^*$ must be an isomorphism, which implies that $L_1 \otimes M^*$ is a field, which contradicts our assumption. Thus (1) implies (2). Since (2) implies the regularity, we have completed the proof. QED

THEOREM 4.5.3 Let L, L' be extension fields of a field K. (1) If L is a regular extension of K, then $L \times_K L'$ is a regular extension of L'. (2) If L, L' are regular extensions of K, then $L \times_K L'$ is a regular extension of K.

PROOF: Let L^*, K^* be the algebraic closures of L', K, respectively. (1) follows from $(L \times_K L') \times_{L'} L^* = L \times_K L^*$, while (2) follows from $(L \times_K L') \times_K K^* = (L \times_K K^*) \times_{K^*} (L' \times_K K^*)$ which is a field by Theorem 4.5.2. QED

COROLLARY 4.5.4 Assume that extension fields L, L' are free over its subfield K. (1) If L is a regular extension of K, then $L \vee L'$ is a regular extension of L'. (2) If L, L' are regular extensions of K, then $L \vee L'$ is a regular extension of K.

PROOF: Easy by virtue of Theorem 4.2.8. QED

4.6 NOETHERIAN RINGS

A module M over a ring R is called noetherian if it satisfies the maximum condition for submodules. If R itself is a noetherian R-module, then we say that R is a noetherian ring.

THEOREM 4.6.1 A module over a ring R is noetherian if and only if every submodule of M is finitely generated. In particular, the ring R is noetherian if and only if every ideal of R is finitely generated.

PROOF: Assume first that M is noetherian, and let N be an arbitrary submodule of M. Let F be the set of finitely generated submodules of N. Since M is noetherian, F has a maximal member, say N^*. If $N^* \neq N$, then let b be an element of N which is not in N^*. Then $N + bR \in F$, which contradicts the maximality of N^*. Thus $N = N^*$ and N is finitely generated. Conversely, assume that every submodule of M is finitely generated. Let $N_1 \subseteq N_2 \subseteq \cdots \subseteq N_r \subseteq \cdots$ be an ascending chain of submodules. Then the union N^* of these N_i is a submodule, which is generated by a finite number of elements, say $a_1, \ldots, a_s$. Then there is a number t such that all the a_i are in N_t, and we see that $N_u = N_t$ for every $u > t$. Thus the ascending chain condition is satisfied, and hence the maximum condition is satisfied. QED

THEOREM 4.6.2. A finitely generated module M over a noetherian ring R is noetherian.

PROOF: Let $a_1, \ldots, a_r$ be a set of generators for M, and we prove the assertion by induction on r. Namely, we assume that every submodule of the module M' generated by $a_2, \ldots, a_r$ is finitely generated. Let N be an arbitrary submodule of M, and let $I = \{b \in R \mid ba_1 \in N + M'\}$. Then I is an ideal of R, hence is generated by a finite number of elements, say $c_1, \ldots, c_s$. Let $f_1, \ldots, f_s$

be elements of N such that $f_i = c_i a_1$ + (some element of M'). We claim that $N = \sum f_i R + (N \cap M')$. Indeed, if $f = b_1 a_1 + \cdots + b_r a_r \in N$ $(b_i \in R)$, then $b_1 \in I$ and $b_1 = \sum e_i c_i$ $(e_i \in R)$. Then $f - \sum e_i f_i \in N \cap M'$, and we see the equality. Now, by our assumption, $N \cap M'$ is finitely generated, and we see that N is finitely generated. This proves the assertion by virtue of Theorem 4.6.1. QED

THEOREM 4.6.3 If a ring R is generated by a finite number of elements over a noetherian ring A, then R is noetherian.

PROOF: It suffices to prove the case where $R = A[x]$ with $x \in R$. Let I be an arbitrary ideal of R. Let J be the set of c $(\in A)$ for which there are elements of I of the form $cx^n + c_1 x^{n-1} + \cdots + c_n$ (for some n and with $c_i \in A$). Then J is an ideal of A, hence generated by a finite number of elements, say $b_1, \ldots, b_r$. Let f_i $(i = 1, \ldots, r)$ be elements of I of the form $b_i x^{n_i}$+(terms of lower degree). Let m be the maximum of $n_1, \ldots, n_r$. We claim that $I = (\sum_i f_i R) + (I \cap (A + Ax + \cdots + Ax^{m-1}))$. Indeed, if $n \geq m$ and if $f = cx^n + \cdots \in I$, then $c \in J$; hence f - (some element of $\sum f_i R$) is of lower degree than f, and we prove the equality by induction on n. Since $I \cap (A + \cdots + Ax^{m-1})$ is a finitely generated A-module by Theorem 4.6.2, we see that the ideal I is finitely generated. QED

An ideal Q of a ring R is called <u>primary</u> if $Q \neq R$ and if every zero-divisor in R/Q is nilpotent; this latter condition is equivalent to: $a, b \in R$, $ab \in Q$, $a \notin Q$ imply $b^n \in Q$ for some natural number n. A prime ideal is primary. For an ideal I of R, $\{c \in R \mid c^n \in I$ for some $n\}$ is called the <u>radical</u> of I and is denoted by $\sqrt{I}$. This is an ideal of R as is easily seen directly or by the following:

THEOREM 4.6.4 The radical $\sqrt{I}$ is the intersection of all prime ideals containing I. If the ring R is noetherian, then $\sqrt{I}$ is <u>nilpotent</u> modulo I (i.e., a power of $\sqrt{I}$ is contained in I).

PROOF: If $a \in \sqrt{I}$ and if P is a prime ideal containing I, then $a^n \in I \subseteq P$ and $a \in P$. Conversely, if b $(\in R)$ is not in $\sqrt{I}$, then the set S of powers of b does not meet I, and therefore there is a maximal ideal P with respect to S such that $I \subseteq P$ and P is a prime ideal not containing b (Theorem 2.4.2). If R is noetherian, then $\sqrt{I}$ is generated by finitely many elements, say $a_1, \ldots, a_s$. $a_i^{n_i} \in I$ for some n_i and $\sqrt{I}^t \subseteq I$ with $t \geq \sum n_i$. QED

THEOREM 4.6.5 If Q is a primary ideal of a ring R, then $\sqrt{Q}$ is prime.

In case $P = \sqrt{Q}$, we say that Q is <u>P-primary</u> or that Q <u>belongs to</u> P.

PROOF: $ab \in \sqrt{Q}$, $a \notin \sqrt{Q}$ imply $a^n b^n \in Q$, $a^n \notin \sqrt{Q}$, and therefore $b^n \in Q$ which means that $b \in \sqrt{Q}$. QED

THEOREM 4.6.6 Assume that Q is an ideal of a noetherian ring R such that $Q \neq R$ and there are no ideals A, B properly containing Q such that $Q = A \cap B$. Then Q is a primary ideal of R.

Before proving the theorem, we introduce the notion of ideal quotients. When I is an ideal of a ring R and S is a subset of R, we denote by $I : S$ the set $\{x \in R \mid xS \subseteq I\}$ which is called the <u>ideal quotient</u> of I with respect to S. Note that $I : S = I : \sum_{s \in S} sR$.

LEMMA 4.6.7 If A, B, C are ideals of a ring R, then

$$(A : B) : C = A : BC,$$

$$(A \cap B) : C = (A : C) \cap (B : C)$$

PROOF: $x \in (A : B) : C$ if and only if $xC \subseteq A : B$, namely $xBC \subseteq A$. $y \in (A \cap B) : C$ if and only if $yC \subseteq A \cap B$, namely $yC \subseteq A$ and $yC \subseteq B$. QED

PROOF of THEOREM 4.6.6: Assume that Q is not primary. Then there are $a, b \in R$ such that $ab \in Q$, $a \notin Q$ and $b^n \notin Q$ $(n = 1, 2, \ldots)$. Set $B_n = Q : b^n R$ $(n = 1, 2, \ldots)$. Since $B_{n+1} = B_n : bR$, we have $B_1 \subseteq B_2 \subseteq \cdots \subseteq B_n \subseteq B_{n+1} \subseteq \cdots$. Since R is noetherian, there is a number N such that $B_N = B_{N+1}$. Set $A = Q + aR$ and $B = Q + b^N R$. By our assumption, A and B contain Q properly. We claim that $Q = A \cap B$. Indeed, if $x \in A \cap B$, then $x = q + ar = q' + b^N r'$ with $q, q' \in Q$, $r, r' \in R$. $bx = qb + abr \in Q$. On the other hand,

$bx = q'b + b^{N+1}r'$ and therefore $r' \in B_{N+1}$ which coincides with B_N by our assumption. Therefore $r' \in B_N$, which shows that $x = q' + b^N r' \in Q$. Thus $Q = A \cap B$. QED

In general, if an ideal A is expressed as the intersection of ideals $B_1, \ldots, B_s$ and if $A \neq B_1 \cap \cdots \cap B_{i-1} \cap B_{i+1} \cap \cdots \cap B_s$ for every $i = 1, \ldots, s$, then we say that $A = B_1 \cap \cdots \cap B_s$ is an irredundant intersection.

LEMMA 4.6.8 Assume that $A = Q \cap Q'$ is an irredundant intersection of primary ideals Q, Q' belonging to prime ideals P, P', respectively. (1) If $P = P'$, then A is P-primary. (2) If $P \neq P'$, then A is not primary.

PROOF: (1): If $x \in P$, then $x^n \in Q$, $x^m \in Q'$, and $x^{n+m} \in A$. Hence $P \subseteq \sqrt{A}$. Since P is prime, we have $P = \sqrt{A}$. If $bc \in A$, $b \notin P$, then $c \in Q$, $c \in Q'$, and $c \in A$. (2): We may assume that $P \not\subseteq P'$. Let b be an element of P which is not in P', and let n be such that $b^n \in Q$. Let c be an element of Q' which is not in Q. Then $b^n c \in QQ' \subseteq A$. But $c \notin A$, $b^n \notin P$. QED

Lemma 4.6.8 shows that if an ideal A is expressed as the intersection of primary ideals $Q_1, \ldots, Q_n$, then we have such an expression that (i) the intersection is irreduntant and (ii) $\sqrt{Q_1}, \ldots, \sqrt{Q_n}$ are all different from each other. An expression satisfying (i) and (ii) is called the shortest representation as an intersection of primary ideals.

THEOREM 4.6.9 In a noetherian ring R, every ideal A is an intersection of a finite number of primary ideals.† If $A = Q_1 \cap \ldots \cap Q_m$ is an irredundant intersection of primary ideals Q_i, then the set $\{\sqrt{Q_i} \mid i = 1, \ldots, m\}$ is uniquely determined by A.

These $\sqrt{Q_i}$ are called associated prime ideals or prime divisors of A.

PROOF: Possibility: Let F be the set of ideals of R which is not an intersection of any finite number of primary ideals. It suffices to show that F is empty. Assume the contrary and let A

---

† R itself is the intersection of primary ideal of number zero.

be a maximal member of F. If $B \cap C = A$ with ideals B, C properly containing A, then, by the maximality of A, $B = Q_1 \cap \cdots \cap Q_m$, $C = Q_{m+1} \cap \cdots \cap Q_n$ with primary ideals Q_i and $A = \cap Q_i$ contradicting $A \in F$. Therefore A is primary by Theorem 4.6.6, which is a contradiction.

Uniqueness: Let $A = Q_1 \cap \cdots \cap Q_m = Q_1' \cap \cdots \cap Q_l'$ be irredundant intersections of primary ideals Q_i, Q_j'. We now prove the equality $\{\sqrt{Q_i} \mid i = 1, \ldots, m\} = \{\sqrt{Q_j'} \mid j = 1, \ldots, n\}$ by induction on $m + n$. Let P be a maximal member in the set of $\sqrt{Q_i}$, $\sqrt{Q_j'}$; we may assume that $P = \sqrt{Q_1}$. Let a be an element of P which is not in any $\sqrt{Q_i}$ or $\sqrt{Q_j'}$ that are different from P. Take a natural number N so that a^N is in all Q_i and Q_j' whose radicals coincide with P. By Lemma 4.6.7, we have $A : a^N = \cap (Q_i : a^N) = \cap (Q_j' : a^N)$. $Q_i : a^N = R$ if $\sqrt{Q_i} = P$, $= Q_i$ otherwise. This is similarly true for Q_j'. Therefore, (i) P must be among $\sqrt{Q_j'}$ (otherwise $A : a^N = A$ and $\cap Q_i$ is redundant), and our induction assumption shows (ii) the required equality except for P. (i) and (ii) show the required equality. QED

Under the above notation a minimal member in the set $\{\sqrt{Q_i} \mid i = 1, \ldots, m\}$ is called a minimal associated prime ideal, or a minimal prime divisor, of A.

THEOREM 4.6.10 If a prime ideal P of a noetherian ring R contains an ideal A, then P contains a prime divisor of A. Therefore a prime ideal P' is a minimal prime divisor of A if and only if it is minimal in the set of prime ideals containing A.

PROOF: Let $A = \cap_{i=1}^{n} Q_i$ be an irredundant intersection of primary ideals Q_i. Then $P \supseteq A \supseteq \Pi Q_i$; hence, for some i, $P \supseteq Q_i$ and $P \supseteq \sqrt{Q_i}$. Thus P contains a prime divisor. QED

We add here a structure theorem of Artin rings:

THEOREM 4.6.11 A ring R satisfies the minimum condition for ideals if and only if (1) R is noetherian and (2) every prime ideal of R is maximal. In this case R is an Artin ring.

PROOF: If part: Assume that (1) and (2) hold. Then every prime ideal must be a minimal prime divisor of 0; hence there are only a finite number of them by virtue of Theorem 4.6.9. Let them be $P_1, \ldots, P_m$ and let $N = \cap P_i$. N is nilpotent (Theorem 4.6.4). Let r be such that $N^r = 0$. By the Chinese remainder theorem (Theorem 2.3.1), we see that R is isomorphic to the direct sum of $R/P_1{}^r, \ldots, R/P_m{}^r$. Each $R/P_i{}^r$ has a chain $R/P_i{}^r \supseteq P_i/P_i{}^r \supseteq \cdots \supseteq P_i{}^r/P_i{}^r = 0$. Since each $P_i{}^s$ is finitely generated, the structure of $P_i{}^s/P_i{}^{s+1}$ as an R-module is that of a finite dimensional vector space over R/P_i. Therefore each $R/P_i{}^r$ has a composition series, and R is an Artin ring.

The only if part: Our proof of Theorem 2.7.8 shows that R has only a finite number of maximal ideals and that every prime ideal P_i of R is maximal. Let N be $\sqrt{0}$, i.e., $N = \cap P_i$. We claim that N is nilpotent. Assume the contrary. Because of the minimum condition, we have a power N^r of N such that $N^r = N^{r+1}$. Let F be the set of ideals B contained in N and such that $BN^r \neq 0$. $N \in F$ and F is not empty. Let B be a minimal member of F, and set $Q = 0 : BN^r$. $Q \neq R$ because $BN^r \neq 0$. Assume that $ab \in Q$, $a \notin Q$ $(a,b \in R)$. Then $aBN^r \neq 0$, and $aB \in F$. Because of the minimality of B, we have $aB = B$. Thus $abBN^r = 0$ implies $bBN^r = 0$ and $b \in Q$. Thus Q is a prime ideal; hence $N \subseteq Q$ and therefore $BN^{r+1} = 0$, which is a contradiction. Thus N is nilpotent and $N^r = 0$. Then $R \cong R/P_1{}^r \oplus \cdots \oplus R/P_m{}^r$. Since each $P_i{}^s/P_i{}^{s+1}$ satisfies the minimum condition and since its structure as an R-module is that of a vector space over R/P_i, we see that each $P_i{}^s/P_i{}^{s+1}$ has a composition series. Therefore R has a composition series and R satisfies the maximum condition for ideals. QED

4.7 RINGS OF QUOTIENTS AND INTEGRAL EXTENSIONS

Let S be a multiplicatively closed subset of a ring R such that $0 \notin S$. Then we define the ring of quotients of R with respect to S, which is denoted by R_S or $R[S^{-1}]$, in the following ways.

(1) As in the case of total quotient rings (Sec. 2.4), consider $Q = \{(a,s) \mid a \in R,\ s \in S\}$. An equivalence relation $\sim$ is defined

by: $(a,s) \sim (b,t)$ if and only if there is u in S such that $bsu = atu$. The equivalence class of (a,s) is denoted by a/s. The set of these classes forms a ring by defining $(a/s) + (b/t) = (at + bs)/st$, $(a/s)(b/t) = ab/st$. This new ring is R_S. For each element a of R, the class as/s is independent of $s \in S$, and we have a mapping ψ of R into R_S such that $\psi a = as/s$. ψ is a ring homomorphism and is called the natural mapping (or homomorphism) of R into R_S.

(2) If S consists only of nonzero-divisors, one may define R_S to be the subring of the total quotient ring of R which is generated by $\{s^{-1} \mid s \in S\}$ over R.

(3) If S contains zero-divisors, then one may define R_S as follows: Set $N = \{a \in R \mid as = 0 \text{ for some } s \in S\}$. Let ψ be the natural homomorphism of R onto R/N. Then, as is easily seen, if s is either an element of S or a nonzero-divisor in R, then ψs is a nonzero-divisor in R/N. Therefore $(R/N)_{\psi S}$ is defined by (2) and is defined to be R_S.

We also define the ring of quotients of a ring R with respect to an ideal A to be R_S with $S = \{a \in R \mid (a \text{ modulo } A) \text{ is a non-zero-divisor in } R/A\}$. This R_S is denoted by R_A. The case where A is a prime ideal (and S is the complement of A) is especially important.

In any case, R_S is an R-module naturally. Hence if T is a subset of R, then TR_S is the ideal of R_S generated by ψT. Since Case (2) is our main case, we use the symbol $\cap$ even in Case (3) in the following way: If I' is an ideal of R_S, $I' \cap R$ denotes $\psi^{-1}(I' \cap \psi R)$.

THEOREM 4.7.1 With R and S as above, (1) if I' is an ideal of R_S, then $I' = (I' \cap R)R_S$, and (2) when Q is a P-primary ideal (P a prime ideal), (i) if $P \cap S$ is not empty then $PR_S = QR_S = R_S$, and (ii) if $P \cap S$ is empty then PR_S is prime, QR_S is PR_S-primary, $QR_S \cap R = Q$, and $PR_S \cap R = P$.

PROOF: Let ψ be the natural mapping of R into R_S. For each s $(\in S)$, ψs is a unit in R_S. If $a/s \in I'$ $(a \in R,\ s \in S)$, then $a \in I' \cap R$ and $a/s = (\psi a)(\psi s)^{-1} \in (I' \cap R)R_S$. Thus we see (1). As for (i), let $s \in S \cap P$. Then a power s^n is in Q and QR_S contains ψs^n which is a unit. Hence (i) is proved. Assume now that $S \cap P$ is empty. If $c \in QR_S \cap R$, then $c = q/s$ with $q \in Q$, $s \in S$. This means that $csu = qu \in Q$ with u in S. Since $su \notin P$, we have $c \in Q$. Thus $QR_S \cap R = Q$. Since P is primary, we have $PR_S \cap R = P$. Assume now that $(a/s)(b/t) \in QR_S$, $b/t \notin QR_S$. Then we have $ab \in Q$, $b \notin Q$. Since Q is primary, we see that a power a^n of a is in Q and $(a/s)^n \in QR_S$. Thus QR_S is primary. If this is applied to P, then the same holds with $n = 1$ and we see that PR_S is prime. $a/s \in PR_S$ implies $a \in P$ and $a^n \in Q$ for some n. Then $(a/s)^n \in QR_S$. Thus (ii) is proved. QED

COROLLARY 4.7.2 Let R and S be as above. If $A = Q_1 \cap \cdots \cap Q_m$ is an irredundant intersection of primary ideals Q_i in the ring R and if Q_i meets S if and only if $i > t$, then $AR_S = Q_1R_S \cap \cdots \cap Q_tR_S$ and this is an irredundant intersection.

PROOF: Assume that $a/s \in \cap_{i \le t} Q_iR_S$ ($a \in R$, $s \in S$). Then $a \in Q_iR_S \cap R = Q_i$ for $i \le t$. Take an $s' \in S \cap (\cap_{i>t} Q_i)$. Then $a/s = as'/ss'$ and $as' \in A$. Thus $AR_S = Q_1R_S \cap \cdots \cap Q_tR_S$. Assume for instance that Q_1R_S is redundant in this intersection. Then $Q_1 \cap \cdots \cap Q_t = (Q_1R_S \cap \cdots \cap Q_tR_S) \cap R = (Q_2R_S \cap \cdots \cap Q_tR_S) \cap R = Q_2 \cap \cdots \cap Q_t$. This contradicts the irredundancy of $Q_1 \cap \cdots \cap Q_m$. QED

The notion of rings of quotients and Theorem 3.1.6 are applied to:

THEOREM 4.7.3 (Lying-over theorem) Assume that a ring R' is integral over its subring R. Let P be a prime ideal of R, and let S be the complement of P in R. Then an ideal P' is a prime ideal of R' such that $P' \cap R = P$ if and only if P' is a maximal ideal with respect to S.

The name came from the term "Lie over" which is defined as follows: When R is a subring of a ring R', and if I, I' are ideals of them such that $I = I' \cap R$, then we say that I' lies over I.

Before proving the theorem, we prove the following lemma.

LEMMA 4.7.4 Assume that a ring R' is integral over its subring R. (1) If ψ is a homomorphism of R' into a ring, then $\psi R'$ is integral over ψR. (2) If S is a multiplicatively closed subset of R such that $0 \notin S$, then R_S is naturally a subring of R'_S and R'_S is integral over R_S.

PROOF: (1) is obvious. As for (2), let ϕ be the natural homomorphism of R' into R'_S. The kernel of ϕ is $N = \{a \in R' \mid as = 0$ for some $s \in S\}$. Therefore $N \cap R$ is the kernel of the natural mapping of R into R_S. Thus we may assume that ϕ is an injection (considering $\phi R'$ and ϕR instead of R' and R) in view of (1) above. Then R'_S is generated by R' over R_S and R' is integral over R; hence every element of R' is integral over R_S. This proves (2). QED

PROOF of THEOREM 4.7.3: Only if part: P' does not meet S, and there is a maximal ideal P^* of R' with respect to S such that $P^* \supseteq P'$. Let ψ be the natural homomorphism of R' onto R'/P', and set $S^* = \psi S$. Then P^* does not meet S^*. Consider $K' = (\psi R')_{S^*}$ and $K = (\psi R)_{S^*}$. Obviously K is a field and K' is an integral domain integral over K (Lemma 4.7.4). Therefore K' must be a field by Theorem 3.1.6. Thus $\psi P^* = 0$, which means $P^* = P'$. Thus a prime ideal lying over P is maximal with respect to S.

If part: Assume that P' is an ideal of R' maximal with respect to S, and set $Q = P' \cap R$. Let ψ be the natural homomorphism of R' onto R'/P', and set $S^* = \psi S$, $K' = (\psi R')_{S^*}$, and $K = (\psi R)_{S^*}$. Since 0 in $\psi R'$ is maximal with respect to S^*, we see that 0 in K' is maximal, and therefore K' is a field. Hence Theorem 3.1.6 shows that K is a field and $P = Q$. QED

COROLLARY 4.7.5 If a ring R' is integral over its subring R, then (1) for every prime ideal P of R, there is a prime ideal P' of R' lying over P; and (2) if P', P'' are prime ideals of R' such that $P' \subset P''$, then $P' \cap R \subset P'' \cap R$.

COROLLARY 4.7.6 (Going-up theorem) Assume that a ring R' is integral over its subring R and $P_1 \subset P_2 \subset \cdots \subset P_n$ is an ascending chain of prime ideals in R. If a prime ideal P_1' of R' lying over P_1 is given, then there is a chain $P_1' \subset P_2' \subset \cdots \subset P_n'$ of prime ideals in R' such that $P_i = P_i' \cap R$ for $i = 2, \ldots, n$.

PROOF: If P_i' is given, then considering R'/P_i' over R/P_i, we see the existence of P_{i+1}' by Corollary 4.7.5. QED

In Corollary 4.7.6, if we give P_n' lying over P_n, then we may not have such a chain of P_i' lying over P_i. We are to give an affirmative answer to the direction under stronger assumption (Theorem 4.7.8 below). For that purpose, we define: Let R be a normal ring with K as its field of fractions, and let L be a normal extension field of K with Galois group $G = G(L/K)$. Let T be the integral closure of R in L. In this case, we say that T is a normal extension of R and that G is its Galois group. (If L is separable over K, then T is called a Galois extension of R.)

THEOREM 4.7.7 Under the circumstances, (1) G is naturally regarded as the set of R-automorphisms of T, and (2) prime ideals P', Q' of T lie over the same prime ideal of R if and only if there is an element σ of G such that $P'^{\sigma} = Q'$.

PROOF: Since conjugates of integral elements are integral, we see that $T^{\sigma} = T$ for every $\sigma \in G$. For $\sigma, \tau \in G$, $\sigma|_T = \tau|_T$ if and only if $\sigma = \tau$, because every element of L is of the form b/c with $b,c \in T$. Every R-automorphism of T is extended to an element of G as is easily seen, and we prove (1). As for (2), the if part is obvious and we assume that P', Q' lie over P. (i) Assume first that $\#(G)$ is finite. Assume that Q' is not a conjugate of P'. Then Lemma 2.4.5 implies the existence of an element $a \in Q'$ which is not in any P'^{σ} $(\sigma \in G)$. Set $b = \Pi_{\sigma \in G} a^{\sigma}$. Then $b^{\sigma} = b$ for any $\sigma \in G$, and b is purely inseparable over K; hence a power b^m of b is in K. Then $Q' \cap R = P$ contains b^m because $a \in Q'$, and on the other hand, $P = P' \cap R$ does not contain b^m because every a^σ is not in P'. This is a contradiction and we have proved this case. (ii) We consider the general case. Let F be the set of pairs $(L_\lambda, \sigma_\lambda)$ of normal extensions L_λ of K and $\sigma_\lambda \in G(L_\lambda/K)$ such that $L_\lambda \subseteq L$, $(P' \cap L_\lambda)^{\sigma_\lambda} = Q' \cap L_\lambda$. Note that $T \cap L_\lambda$ is the integral closure of R in L. We introduce an order in F by: $(L_\lambda, \sigma_\lambda) \geq (L_\mu, \sigma_\mu)$ if and only if $L_\lambda \supseteq L_\mu$ and $\sigma_\lambda|_{L_\mu} = \sigma_\mu$. Then F becomes an inductive set and F has a maximal element, say (L^*, σ^*). Assume that $L^* \neq L$, and let a be an element of L which is not in L^*. Since the number of conjugates of a is finite, the least normal extension L' containing $L^*(a)$ is finite over L^*. σ^* extends to an automorphism of the algebraic closure of L. Then, since L is normal, σ^* is regarded as an element of G. Then $(P' \cap L')^{\sigma^*}$ and $Q' \cap L'$ are prime ideals of $T' = T \cap L'$ lying over $Q' \cap L^*$. By the case (i), there is $\tau \in G(L'/L^*)$ such that $(P' \cap L')^{\sigma^*\tau} = Q' \cap L'$ which shows that $(L', \sigma^*\tau) \in F$, contradicting the maximality of (L^*, σ^*). Thus $L^* = L$. QED

THEOREM 4.7.8 (Going-down theorem) Assume that an integral domain T is integral over its subring R which is normal. If an ascending chain $P_1 \subset P_2 \subset \cdots \subset P_n$ of prime ideals in R and a prime ideal P'_n of T lying over P_n are given, then there is a chain $P'_1 \subset P'_2 \subset \cdots \subset P'_n$ of prime ideals in T such that $P'_i \cap R = P_i$.

PROOF: (1) Assume first that T is normal over R. Let $P''_1 \subset P''_2 \subset \cdots \subset P''_n$ be an ascending chain of prime ideals in T such that $P''_i \cap R = P_i$ (going-up theorem). Let $\sigma \in G(T/R)$ be such that ${P''_n}^\sigma = P'_n$. Then $P'_i = {P''_i}^\sigma$ are the required prime ideals. (2) In the general case, let T^* be a normal extension of R containing T, and let P^*_n be its prime ideal lying over P'_n. Then by (1) we have $P^*_1 \subset \cdots \subset P^*_n$ in this case, and $P'_i = P^*_i \cap T$ are the required. QED

4.8 KRULL DIMENSION

Let I be an ideal of a ring R such that $I \neq R$. Then the height of I, which is denoted by ht I, is defined as follows: (1) When I is a prime ideal, ht $I = n < \infty$ means that n is the length of the longest descending chain of prime ideals in R which begins with I; otherwise ht I = ∞. (2) In the general case, ht I is the minimum of ht P where P runs through prime ideals containing I. The <u>Krull dimension</u> of R, which is denoted by Krull dim R, is defined to be the maximum of ht P, where P runs through prime ideals of R; if the maximum does not exist, then Krull dim R is defined to be ∞.

THEOREM 4.8.1 If a ring T is integral over its subring R, then Krull dim T = Krull dim R.

PROOF: Corollary 4.7.6 shows that Krull dim T $\geq$ Krull dim R. If $P_0 \supset \cdots \supset P_n$ is a descending chain of prime ideals in T, then $P_0 \cap R \supset \cdots \supset P_n \cap R$ by Corollary 4.7.5. QED

Note that this result does not imply that if a prime ideal P' of T lies over a prime ideal P of R then ht P' = ht P, even if T is an integral domain (and one can have a counterexample; see for instance, pp.203-205 in, Nagata, <u>Local rings</u>, John Wiley, New York, 1962; reprint ed., Krieger, Huntington, N.Y.). But we have:

THEOREM 4.8.2 Assume that an integral domain T is integral over its subring R which is normal. Then for an arbitrary ideal I of T such that $I \neq T$, we have ht I = ht $(I \cap R)$.

PROOF: (1) Assume first that I is prime. If $I = P_0 \supset \cdots \supset P_n$ is a chain of prime ideals in T, then $P_0 \cap R \supset \cdots \supset P_n \cap R$ by Corollary 4.7.5, and ht I $\leq$ ht $(I \cap R)$. Conversely, if $I \cap R = Q_0 \cap \cdots \cap Q_m$ is a chain of prime ideals in R, then Theorem 4.7.8 shows the existence of a chain of prime ideals $I = P_0 \supset \cdots \supset P_m$ in T such that $P_i \cap R = Q_i$. Thus ht I = ht $(I \cap R)$ in this case. (2) General case: ht I = ht P with a prime ideal P containing I; $P \cap R \supseteq I \cap R$ and ht P = ht $(P \cap R)$. Thus ht I $\geq$ ht $(I \cap R)$. On the other hand, let Q be a prime ideal of R such that $Q \supseteq I \cap R$ and ht Q = ht $(I \cap R)$. By the lying-over theorem applied to T/I over $R/(I \cap R)$, we see the existence of a prime ideal Q' of T containing I and lying over Q. Then ht Q' = ht Q, and ht I $\leq$ ht $(I \cap R)$. QED

4.9 NORMALIZATION THEOREMS

LEMMA 4.9.1 If f is in the polynomial ring $F = K[X_1,\dots,X_n]$ in variables $X_1, \dots, X_n$ over a field K and if $f \notin K$, then there are elements $Y_1, Y_2, \dots, Y_n$ of F such that (i) $Y_1 = f$ and (ii) F is integral over $R = K[Y_1,\dots,Y_n]$. (1) We can choose Y_i so that $Y_i = X_i + X_1^{m_i}$ $(i = 2, \dots, n)$; if q is a given natural number, then all m_i can be chosen to be multiples of q. (2) If $\#(K) = \infty$, then we can choose Y_i so that $Y_i = X_i + c_i X_1$ with $c_i \in K$ for $i = 2, \dots, n$.

PROOF: Let t be a multiple of q such that $t > \deg f$ and set $m_i = t^{i-1}$ and $Y_i = X_i + X_1^{m_i}$ for $i \geq 2$; $m_1 = 1$. For each monomial $M = X_1^{d_1}\cdots X_n^{d_n}$, we define its weight $w(M)$ to be $\sum d_i m_i$. Then, among monomials M, $M' = X_1^{d_1'}\cdots X_n^{d_n'}$ of degrees less than t, $w(M) > w(M')$ if and only if $(d_n,\dots,d_1) > (d_n',\dots,d_1')$ in the lexicographical order. Therefore, among monomials appearing in f, there is an M such that $w(M)$ is the biggest. Then, if we write f as a polynomial in $X_1, Y_2, \dots, Y_n$, then we see that $f = cX_1^{w(M)} +$ (terms of lower degree in X_1) with $c =$ (the coefficient of M in f). This shows that X_1 is integral over $R = K[Y_1, Y_2,\dots,Y_n]$. Since $X_i = Y_i - X_1^{m_i}$, we see that F is integral over R. If $\#(K) = \infty$, then we set $Y_i = X_i + c_i X_1$ and express f as a polynomial in $X_1, Y_2, \dots, Y_n$. The coefficient a of $X_1^{\deg f}$ in the new expression is a polynomial in the c_i. Since $\#(K) = \infty$, we can choose these c_i so that $a \neq 0$, and we prove (2). QED

THEOREM 4.9.2 (<u>Normalization theorem for polynomial rings</u>) Let I be an ideal of the polynomial ring $F = K[X_1,\dots,X_n]$ over a field K, and let P be the prime field. Set $h = \operatorname{ht} I$. Then there are elements $Y_1, \dots, Y_n$ of F such that (1) F is integral over $R = K[Y_1,\dots,Y_n]$, (2) $I \cap R$ is generated by $Y_1, \dots, Y_h$, and (3) $Y_{h+i} = X_{h+i} + g_i$ with $g_i \in P[X_1,\dots, X_h]$; in case $p =$ (the characteristic of K) $\neq 0$, $g_i \in P[X_1^p,\dots,X_h^p]$.

PROOF: We employ induction on h. If $h = 0$, then $I = 0$, and we may set $X_i = Y_i$. Assume that $h \geq 1$. Take an ideal I' such that $I' \subset I$, ht $I' = h - 1$. Then there are $Y_1', \ldots, Y_n'$ satisfying corresponding conditions for I'. Set $R' = K[Y_1',\ldots,Y_n']$. Then $I' \cap R' = \sum_{i<h} Y_i'R'$. Theorem 4.8.2 shows that $\mathrm{ht}(I' \cap R') = h - 1$, $\mathrm{ht}(I \cap R') = h$ (Theorem 2.9.5). Therefore, there is an element $f \in I \cap R'$ such that $f \notin I' \cap R'$. Considering $f(0,\ldots,0,Y_h',\ldots,Y_n')$, we may assume that $f \in F' = K[Y_h',\ldots,Y_n']$. Apply Lemma 4.9.1 to this f and F', and we see the existence of $Y_h = f, Y_{h+1}, \ldots, Y_n$ such that F' is integral over $K[Y_h,\ldots,Y_n]$ and $Y_{h+i} = Y_{h+i}' + Y_h'^{m_i}$ ($i \geq 1$). Set $Y_i = Y_i'$ for $i < h$. Then, with $Y_1, \ldots, Y_n$, we see (1) and (3) immediately (in case $p \neq 0$, we choose m_i to be multiples of p). As for (2), since $I \cap R$ contains $\sum_{i\leq h} Y_iR$ which is a prime ideal of height at least h, we see that $I \cap R = \sum_{i\leq h} Y_iR$. QED

COROLLARY 4.9.3 (<u>Normalization theorem for finitely generated ring; Normalization theorem of Noether</u>) If a ring R is generated by elements $a_1, \ldots, a_n$ over a field K, then there are elements $z_1, \ldots, z_t$ of R such that (1) R is integral over $K[z_1,\ldots,z_t]$ and (2) $z_1, \ldots, z_t$ are algebraically independent over K.

PROOF: Let I be the kernel of the natural surjection ψ of the polynomial ring $K[X_1,\ldots,X_n]$ to R. Then we have $Y_1, \ldots, Y_n$ as in Theorem 4.9.2. Set $z_i = \psi Y_{h+i}$ ($i = 1, \ldots, n - h$). Then $R = \psi(K[X_1,\ldots,X_n])$ is integral over $\psi(K[Y_1,\ldots,Y_n]) = K[z_1,\ldots,z_{n-h}]$. If $g \in K[Z_1,\ldots,Z_{n-h}]$ (polynomial ring) and if $g(z_1,\ldots,z_{n-h}) = 0$, then $g(Y_{h+1},\ldots,Y_n) \in I$, and $g = 0$ by (2) in Theorem 4.9.2. QED

We give some applications of these normalization theorems.

THEOREM 4.9.4 Assume that an integral domain R is generated by a finite number of elements over a field K. If $P_0 \supset P_1 \supset \cdots \supset P_n = 0$ is a chain of prime ideals such that (i) P_0 is maximal and (ii) this chain has no refinement as a chain of prime ideals, then $n = \mathrm{trans.deg}_K R$.

PROOF: We prove this by induction on n. Let $z_1, \ldots, z_t$ be as in Corollary 4.9.3. Then $t = \text{trans.deg}_K R$, and R is integral over $F = K[z_1,\ldots,z_t]$. If $n = 0$, then R is a field, and F is a field by Theorem 3.1.6. Then $t = 0 = n$. Assume now that $n \geq 1$. Applying Lemma 4.9.1 to $P_{n-1} \cap F$ and F, we may assume that $z_1 \in P_{n-1}$. By our induction assumption applied to R/P_{n-1} over $K[z_2, \ldots, z_t]$, we see that $n - 1 = t - 1$. QED

COROLLARY 4.9.5 Assume that an integral domain R is generated by a finite number of elements over a field K. Then for every prime ideal P of R, it holds that $\text{trans.deg}_K R = \text{trans.deg}_K R/P + \text{ht } P$. In particular, if M is a maximal ideal of R, then R/M is algebraic over K.

Corollary 4.9.6 (Hilbert zero-point theorem) In a ring R generated by finitely many elements over a field K, the radical of an ideal I coincides with the intersection of maximal ideals containing I.

PROOF: Assume that $f \notin \sqrt{I}$ $(f \in R)$, and it suffices to show the existence of a maximal ideal containing I but not f. There is a prime ideal P containing I but not f. Then we may replace I with P. Then, considering R/I instead of R, we may assume that $I = 0$ and that R is an integral domain. Consider $R' = R[f^{-1}]$. Let M' be a maximal ideal of R'. Then R'/M' is algebraic over K (Corollary 4.9.5); hence $R/(M' \cap R)$ is algebraic over K. This shows that $M' \cap R$ is maximal in R. Since $f^{-1} \in R'$, $f \notin M' \cap R$. QED

4.10 INTEGRAL CLOSURES

LEMMA 4.10.1 Let R be a noetherian ring with total quotient ring Q. An R-submodule M of Q is finitely generated if and only if there is a nonzero-divisor a of R such that $aM \subseteq R$.

PROOF: If part: M is contained in $a^{-1}R$ which is a finite R-module, and hence M is noetherian (Theroem 4.6.1). The only if part is easy. QED

THEOREM 4.10.2 Let R be a normal ring, $f(X)$ a monic polynomial over R, a a root of $f(X)$ in an extension field, and R^*

the derived normal ring of $R[a]$. Let $f'(X)$ be the derivative of $f(X)$, and let D be the discriminant of $f(X)$. Then we have $DR^* \subseteq f'(a)R^* \subseteq R[a]$.

PROOF: Let the roots of $f(X)$ be $u_1 = a, u_2, \ldots, u_r$ ($r = \deg f$), and set $g_i(X) = f(X)/(X - u_i)$. Then $f'(X) = \sum g_i(X)$ and $f'(a) = g_1(a) = \Pi_{i>1} (u_1 - u_i)$. Therefore $DR^* \subseteq f'(a)R^*$. It suffices to prove the inclusion $f'(a)R^* \subseteq R[a]$ assuming that $f(X)$ is irreducible. If a is inseparable, then $f'(a) = 0$. Hence we assume that a is separable. Let T be the least Galois extension of R containing R^*, let G be the Galois group, and let H be the subgroup corresponding to R^*. Let $\sigma_1 = 1, \sigma_2, \ldots, \sigma_r$ be elements of G such that $a^{\sigma_i} = u_i$. Then G is the disjoint union of $H\sigma_i$. $g_i(X) = g_1(X)^{\sigma_i}$. Write $g_1(X) = c_{r-1}X^{r-1} + \cdots + c_0$ ($c_{r-1} = 1$; $c_i \in R[a]$). Let b be an arbitrary element of R^*. Then $bf'(a) = bg_1(a) = \sum_{i=1}^r b^{\sigma_i} g_i(a) = \sum_{i,j} b^{\sigma_i} c_j^{\sigma_i} a^j = \sum_j (\sum_i b^{\sigma_i} c_j^{\sigma_i}) a^j$. Since b, c_j are H-invariants, we see that $\sum_i b^{\sigma_i} c_j^{\sigma_i}$ is G-invariant and hence is in R. Thus $bf'(a) \in R[a]$. QED

COROLLARY 4.10.3 If R is a noetherian normal ring, K its field of fractions, and let L be a finite separably algebraic extension field of K. If a ring R', such that $R \subseteq R' \subseteq L$, is integral over R, then R' is finite as an R-module.

PROOF: Let $a \in L$ be such that $L = K(a)$ and let $c_0, \ldots, c_n$ R be such that $c_0 a^n + \cdots + c_n = 0$ ($c_0 \neq 0$). Then, taking $c_0 a$ instead of a, we may assume that a is integral over R. Then Lemma 4.10.1 and Theorem 4.10.2 show that the derived normal ring R^* of $R[a]$ is a finite $R[a]$-module, hence over R, too. Thus R' is a noetherian R-module. QED

LEMMA 4.10.4 In Corollary 4.9.3, if R is an integral domain which is <u>separable</u> over K (i.e., the field of fractions of R is separable over K), then these $z_1, \ldots, z_t$ can be chosen so that R is separably algebraic (and integral) over $K[z_1,\ldots,z_t]$.

PROOF: It sufficies to prove the case where the characteristic $p \neq 0$. Let $a_1, \ldots, a_n$ be a set of generators for R over K such that $a_{h+1}, \ldots, a_n$ form a separating transcendence base. Then we apply the proof of Corollary 4.9.3. By our construction of Y_i in Theorem 4.9.2, we see that $a_{r+i} - z_i \in P[a_1^p,\dots,a_h^p]$. Therefore, for every derivation D of $K(a_1,\dots,a_n)$, we have $Da_{h+i} = Dz_i$ $(i = 1, \dots, n - h;\ n - h = t.)$. This shows that if D is a derivation over $K(z_1,\dots,z_t)$, then D is a derivation over $K(a_{h+1},\dots, a_n)$. Since $a_{h+1}, \dots, a_n$ form a separating transcendence base, it follows that $D = 0$, and $z_1, \dots, z_t$ are the required (Theorem 4.4.2). QED

Now we come to another application of the normalization theorems.

THEOREM 4.10.5 Let R be a finitely generated integral domain over a field K, and let L be the field of fractions of R. Then the integral closure R^* of R in a finite algebraic extension field L' of L is a finite R-module, and hence R^* is finitely generated over K.

PROOF: If L' is separably generated over K, then we choose $z_1, \dots, z_t$ as in Lemma 4.10.4, and we prove the assertion by Corollary 4.10.3. In the general case, we choose $z_1, \dots, z_t$ as in Corollary 4.9.3. By Corollary 4.4.5, there is a finite purely inseparable extension K' of K such that $L' \vee K'$ is separable over K'. Then the integral closure T^* of $T = K'[z_1,\dots,z_t]$ is a finite T-module. T is a finite module over $F = K[z_1,\dots,z_t]$, and T^* is finite over F. Since $F \subseteq R \subseteq R^* \subseteq T^*$, we see the result. QED

COROLLARY 4.10.6 Let R be a finitely generated integral domain over a field K, and let R^* be the derived normal ring of R. Let P^* be a prime ideal of R^*, and set $P = P^* \cap R$. Then $\operatorname{ht} P^* = \operatorname{ht} P$.

PROOF: Since R^* is finitely generated, we have $\operatorname{ht} P^* = \operatorname{trans.deg}_K R^* - \operatorname{trans.deg}_K R^*/P^* = \operatorname{trans.deg}_K R - \operatorname{trans.deg}_K R/P = \operatorname{ht} P$ (Corollary 4.9.5). QED

Although the following theorem is known to general noetherian rings, we prove this restricted form as an application of the normalization theorems.

THEOREM 4.10.7 If a ring R is finitely generated over a field K and if I is an ideal of R generated by r elements $(I \neq R)$, then for every minimal prime divisor P of I, it holds that $\operatorname{ht} P \leq r$.

PROOF: (1) The case where $r = 1$: It suffices to show a contradiction assuming the existence of a chain of prime ideals $P \supset P'$ $\supset P''$. Considering R/P'', we may assume that R is an integral domain. By Corollary 4.10.6, we may assume that R is normal. Let $Q_1, \ldots, Q_m$ be minimal prime divisors of I different from P, and let g be an element of $Q_1 \cap \cdots \cap Q_m$ outside of P. Considering $R[g^{-1}]$, we may assume that P is the unique minimal prime divisor of I (cf. Theorem 4.7.1). Let $z_1, \ldots, z_t$ be as in Corollary 4.9.3. Set $F = K[z_1,\ldots,z_t]$. Let L, M be fields of fractions of R, F, respectively, and consider a finite normal extension L^* of M containing L, the integral closure R^* of F in L^*, the norm $f^* = N_{L/M}(f)$ of a generator f of I (cf. Exercise 3.6.3), and a prime ideal P^* of R^* lying over P. P^* is a minimal prime divisor of fR^* (Corollary 4.7.5). Since F is a UFD, there is a prime ideal Q of F containing f^*, contained in $P \cap F$ and of height 1. By the going-down theorem, there is a prime ideal Q^* of R^* lying over Q and contained in P^*. Then $f^* \in Q^*$, and some conjugate of f is in Q^*. Then $f \in Q^{*\sigma}$ with a σ in $G(R^*/F)$. By Theorem 4.8.2, we have $\text{ht}\ (Q^{*\sigma} \cap R) = \text{ht}\ Q^{*\sigma} = \text{ht}\ Q = 1$, and therefore $Q^{*\sigma} \cap R$ is a minimal prime divisor of fR. This implies that $Q^{*\sigma} \cap R = P$ and $\text{ht}\ P = 1$.

(2) The general case: We prove the assertion by induction on r. Let $f_1, \ldots, f_r$ be a set of generators for I, and set $I' = \sum_{i<r} f_iR$. Let P' be a minimal prime divisor of I' contained in P. By induction, $\text{ht}\ P' \le r - 1$. In R/P', P/P' is a minimal prime divisor of the principal ideal $f_rR + P'/P'$ and $\text{ht}\ P/P' \le 1$. Therefore, by virtue of Corollary 4.9.5, we see that $\text{ht}\ P \le r$. QED

4.11 CONDITION C_i

Let $f_1(X_1,\ldots,X_n), \ldots, f_r(X_1,\ldots,X_n)$ be polynomials in $X_1, \ldots, X_n$ over a field K such that every f_i has no constant term. Then the equation $f_1 = \cdots = f_r = 0$ has a solution $(0,\ldots,0)$, which is called the <u>trivial</u> solution. Existence of a nontrivial solution in a given field K is usually difficult, and we want to discuss some existence theorems.

We say that K satisfies the condition C_i or C_i' (i is a nonnegative real number) or that K is a C_i-field or C_i'-field, respectively, if the respective condition as follows is satisfied by the field K:

(C_i) If $f(X_1,\ldots,X_n)$ is a homogeneous form of degree d over K and if $n > d^i$, then $f = 0$ has a nontrivial solution in K.

(C_i') If $f(K_1,\ldots,X_n)$ is a polynomial of degree d over K having no constant term and if $n > d^i$, then $f = 0$ has a nontrivial solution in K.

Obviously, C_i' is stronger than C_i.

THEOREM 4.11.1 If K is a C_i-field with $i < 1$, then K is algebraically closed, and conversely.

PROOF: Assume that K is not algebraically closed, and let L be a finite normal extension of K. Set $G = G(L/K)$, $e = [L : K]_i$. Taking a linearly independent base $a_1, \ldots, a_n$ of L over K, we set $g(X_1,\ldots,X_n) = \Pi_{\sigma\in G} (\sum_j a_j^{\sigma} X_j)^e$. Then g is a homogeneous form over K of degree n in n variables. If $(b_1,\ldots,b_n)$ is a solution of $g = 0$ in K, then $\sum a_j^{\sigma} b_j = 0$ for some $\sigma \in G$. Since $b_j \in K$, it follows that $\sum a_j b_j = 0$ and $b_j = 0$ by the linear independence of the a_j. Thus K cannot be a C_i-field for any $i < 1$. The converse is obvious. QED

In order to deal with C_i and C_i' simultaneously, we mean by a form of degree d only in this section either a homogeneous form of degree d or a polynomial having no constant term and of degree at most d, according to the C_i case or the C_i' case, respectively.

THEOREM 4.11.2 Assume that K is a C_i-field or a C_i'-field. If $f_1, \ldots, f_r$ are forms of degree d in $X_1, \ldots, X_n$ over K and if $n > rd^i$, then $f_1 = \cdots = f_r = 0$ has a nontrivial solution in K.

In order to prove this theorem, we introduce a notation: If g is a form in $X_1^{(0)}, \ldots, X_{N_0}^{(0)}$ and if $f_1, \ldots, f_r$ are forms of degree d in $X_1, \ldots, X_n$, then by $g^{(j)}(f_1,\ldots,f_r)$ we understand the form in $X_1^{(j)}, \ldots, X_{N_j}^{(j)}$ defined as follows:

$g^{(0)}(f_1,\ldots f_r) = g(X_1{}^{(0)},\ldots, X_{N_0}{}^{(0)})$ (i.e., g itself).

If $g^{(s)}(f_1,\ldots,f_r) = g_s(X_1{}^{(s)}, \ldots, X_{N_s}{}^{(s)})$ is defined, and t_s is the integer part of N_s/r, we set $N_{s+1} = t_sn$ and we define $g^{(s+1)}(f_1,\ldots,f_r)$ to be the polynomial obtained from g_s replacing $X_j{}^{(s)}$ with $f_m(X_{qn+1}{}^{(s+1)}, \ldots, X_{qn+n}{}^{(s+1)})$ if $j = qr + m$, $q < t_s$; with zero otherwise (i.e., if $j > t_sr$). Under the circumstances, we have the following two results.

LEMMA 4.11.3 If both $f_1 = \cdots = f_r = 0$ and $g = 0$ have no nontrivial solution in K, then for every natural number s, $g^{(s)}(f_1, \ldots,f_r) = 0$ has no nontrivial solution in K.

LEMMA 4.11.4 If m is a nonnegative real number and if $N_0 \geq r$ and $n > rd^m$, then, letting D_s be the degree of $g^{(s)}(f_1,\ldots,f_r)$, we have

$$\lim_{s\to\infty} N_s/D_s{}^m = \infty$$

PROOF of LEMMA 4.11.3: It suffices to prove the case $s = 1$. Assume that $(a_1,\ldots,a_N)$ $(N = N_1)$ is a solution of $g^{(1)}(f_1,\ldots,f_r) = 0$, namely, of $g_1(X_1{}^{(1)},\ldots,X_N{}^{(1)}) = 0$, in K. Since $g = 0$ has no nontrivial solution and since $(f_1(a_1,\ldots,a_n), f_2(a_1,\ldots,a_n), \ldots)$ is a solution of $g = 0$, we have $f_1(a_1,\ldots,a_n) = \cdots = f_r(a_1,\ldots,a_n) = 0$. By our assumption, we have $(a_1,\ldots,a_n) = (0,\ldots,0)$. Similarly, $(a_{wn+1},\ldots,a_{wn+n}) = (0,\ldots,0)$ for each w. QED

PROOF of LEMMA 4.11.4: Obviously $D_{s+1} \leq dD_s$. Set $b = n - rd^m$. Since $N_{s+1} = nt_s$ $[t_sr \leq N_s < (t_s + 1)r]$, we have $N_{s+1}/D_{s+1}{}^m \geq nt_s/d^mD_s{}^m = (rd^m + b)t_s/d^mD_s{}^m = rt_s(1 + (b/rd^m))/D_s{}^m = (N_s - u_s)(1 + (b/rd^m))/D_s{}^m$ with $u_s < r$. Therefore $N_{s+1}/D_{s+1}{}^m - N_s/D_{s+1}{}^m \geq (b/rd^m)(N_s/D_s{}^m) - u_s(1 + (b/rd^m))D_s{}^m$. Since b/rd^m is a positive constant, since N_s tends to infinity, and since $u_s < r$, we see that $\lim N_s/D_s{}^m = \infty$. QED

PROOF of THEOREM 4.11.2: If K is algebraically closed, then the assertion is obvious. Assume that K is not algebraically closed. Then there is a form $g(X_1,\ldots,X_N)$ of degree N (>1) such that

$g = 0$ has no nontrivial solution in K by the proof of Theorem 4.11.1. Then, considering $g^{(s)}(g)$, we may assume that $N > r$ (Lemma 4.11.3, Lemma 4.11.4). Then, we consider $g^{(s)}(f_1,\ldots,f_r)$ such that $N_s/D_s{}^i > 1$ (Lemma 4.11.4). Since K is either a C_i-field or a C_i'-field, $g^{(s)}(f_1,\ldots,f_r) = 0$ has a nontrivial solution in K; hence $f_1 = \cdots = f_r = 0$ has a nontrivial solution in K by Lemma 4.11.3. QED

THEOREM 4.11.5 Assume that K is either a C_i-field or a C_i'-field. If an extension field L of K has transcendence degree t over K, then L is either a C_{i+t}-field or a C_{i+t}'-field, respectively.

PROOF: Let $g(X_1,\ldots,X_n)$ be a form over L of degree d such that $n > d^{i+t}$. It suffices to prove that $g = 0$ has a nontrivial solution in L. Therefore we may replace L with the field generated by the coefficients of g over K. Thus we may assume that L is finitely generated over K.

(1) The case where $t = 0$: Take $g_s = g^{(s)}(g)$ so that $N_s/D_s{}^i > [L : K]^i$. Consider the norm $f = N_{L/K}(g_s)$. Then f is a form in N_s variables over K and of degree $D_s \times [L : K]$. By our choice of s, $f = 0$ has a nontrivial solution in K, and therefore $g = 0$ has a nontrivial solution in L.

(2) The case where $t > 0$: It suffices to prove the case where $L = K(z)$ with a transcendental element z [by induction on t and by (1) above]. We may assume that the coefficients of g are in $K[z]$. Let r be the maximum of the degrees in z of the coefficients of g. Since $n > d^{i+1}$, there is a natural number m such that $m(n - d^{i+1}) > d^i(r+1) - n$. We introduce new variables Y_{jk} ($j = 1, \ldots, n$; $k = 0, 1, \ldots, m$) and set $X_j = \sum_{k=0}^m Y_{jk}z^k$. Then $g = f_0 + f_1 z + \cdots + f_{dm+r}z^{dm+r}$ with forms f_i in those Y_{jk} over K of degree d. Since $n(m + 1) > d^i(dm + r + 1)$, $f_0 = \cdots = f_{dm+r} = 0$ has a nontrivial solution (y_{jk}) in K by Theorem 4.11.2. Then $x_i = \sum y_{jk}z^k$ give a nontrivial solution of $g = 0$ in $K(z)$. QED

THEOREM 4.11.6 A finite field is a C_1'-field.

This follows from the following:

THEOREM 4.11.7 Let K be a finite field of characteristic p, and set $q = \#(K)$. If a polynomial $f(X_1,\ldots,X_n)$ over K is of degree d and if $n > d$, then the number N of solutions of $f = 0$ in K is a multiple of p.

PROOF: Set $S = \{(a_1,\ldots,a_n) \mid a_i \in K\}$. Then $\#(S) = q^n$. $0 \neq x \in K$ implies $x^{q-1} = 1$. Therefore $\sum_{a\in S} f(a)^{q-1} = q^n - N = -N$ (in K). Thus it suffices to show that $\sum_{a\in S} f(a)^{q-1} = 0$ for every polynomial f of degree $\leq d$, or, that $\sum_{a\in S} g(a) = 0$ for every monomial g of degree at most $d(q - 1)$. Let $g = X_1^{e_1}\cdots X_n^{e_n}$. $\sum g(a) = \sum_{c_j\in K} (\Pi_j c_j^{e_j}) = \sum c_1^{e_1}(\Pi_{j\geq 2} c_j^{e_1}) = \sum (\sum c_1^{e_1})(\Pi_{j\geq 2} c_j^{e_j})$. Similarly, we have $\sum g(a) = \Pi_j(\sum_{c\in K} c^{e_j})$. If $e = 0$, then $\sum_{c\in K} c^e = 0$. Assume that e is a natural number which is not a multiple of $q - 1$. Let s be the largest common divisor of e and $q - 1$, and set $t = (q - 1)/s$. Let u be a generator of the cyclic group $K - \{0\}$. Then $\sum_{c\in K} c^e = s(1 + u^e + \cdots + u^{(t-1)e}) = s(1 - u^{te})/(1 - u^e) = 0$. Since $\deg g = \sum e_j \leq d(q - 1)$, e_j in some factor $\sum c^{e_j}$ must be one of the two cases of e above, and we have $\sum g(a) = 0$. QED

4.12 THEOREM OF LÜROTH

LEMMA 4.12.1 Let x be a transcendental element over a field K. If $y = f(x)/g(x)$ [$f(x)$, $g(x) \in K[x]$ and f, g have no proper common divisor], then the polynomial $f(X) - g(X)y$ over $K(y)$ is irreducible. Hence $[K(x) : K(y)] = \max \{\deg f, \deg g\}$.

PROOF: $f(X) - g(X)y$ is irreducible as a polynomial in y over $K(X)$. Since $K[X]$ is a UFD, $f(X) - g(X)y$ is irreducible as a polynomial in X, y over K (Corollary 2.6.6). QED

THEOREM 4.12.2 (Theorem of Lüroth) If $x_1, \ldots, x_r$ are algebraically independent over a field K and if L is a field such that $K \subset L \subseteq K(x_1,\ldots,x_r)$ and $\text{trans.deg}_K L = 1$, then $L = K(z)$ with a transcendental element z.

PROOF: (1) The case where $r = 1$: Let $h(X)$ be the minimal polynomial for $x = x_1$ over L. Multiplying a suitable element of

$K[x]$, we obtain from $h(x)$ a primitive polynomial $h^*(X,x) = d_0(x)X^n + d_1(x)X^{n-1} + \cdots + d_n(x)$ over $K[x]$. Let y be one of d_i/d_0 which is not in K, and write $y = f(x)/g(x)$ as in Lemma 4.12.1. Then the minimal polynomial for x over $K(y)$ is an associate to $f(X) - g(X)y$. Since $h(X)$ is the minimal polynomial for x over L and since $y \in L$, $h(X)$ divides $f(X) - g(X)y$ as a polynomial over L: $f(X) - g(X)y = h(X)q(X)$. Since $h^*(X,x)$ is a primitive polynomial over $K[x]$, we have $f(X)g(x) - g(X)f(x) = h^*(X,x)q^*(X,x)$. The degree in x on the left-hand side is the maximum of deg f and deg g, while the one on the right-hand side is at least the maximum of deg d_i $(i = 0, \ldots, n)$. This shows that $q^*(X,x)$ is a polynomial only in X. If q^* is not a constant, then taking a root c, we have $f(c)g(x) - g(c)f(x) = 0$, which is a contradiction. Thus $q^* \in K$. Thus $[K(x) : L] = [K(x) : K(y)]$, and $L = K(y)$.

(2) The case where $r \geq 2$: We employ the following lemma, which will be proved in Sec. 5.12 [cf. (3) below]:

LEMMA 4.12.3 Let L be an extension field of a field K, and let x be transcendental over L. If x, y are algebraically independent over K and if $L(x) = K(x,y)$, then $L \cong K(y)$.

Now, we use an induction on r. We may assume that x_1 is transcendental over L. $K(x_1) \subset L(x_1) \subseteq K(x_1,\ldots,x_r)$ and therefore $L(x_1) = K(x_1, y)$ by our induction. Thus we see the result by our lemma.

(3) The case where $\#(K) = \infty$ can be proved in a more general form as follows:

LEMMA 4.12.4 Let L, M be finitely generated fields over an infinite field K. If there is an element x which is transcendental over both L and M such that $L(x) = M(x)$, then L and M are K-isomorphic to each other.

PROOF: Let $L = K(y_1,\ldots,y_s)$, $M = K(z_1,\ldots,z_t)$. There are rational expressions $y_i = f_i(z_1,\ldots,z_t,x)$, $z_j = g_j(y_1,\ldots,y_s,x)$ $(i = 1, \ldots, s;\ j = 1, \ldots, t)$. Let a be an element of K such that none of the denominators of f_i, g_j vanishes when we put a

in place of x. Set $b_i = f_i(z_1,\ldots,z_t,a)$, $c_j = g_j(y_1,\ldots,y_s,a)$. Since $y_i = f_i(g_1,\ldots,g_t,x)$ and since x is transcendental over $L = K(y_1,\ldots,y_s)$, we have $y_i = f_i(c_1,\ldots,c_t,a)$. Similarly, $z_j = g_j(b_1,\ldots,b_s,a)$. Thus $L = K(y_1,\ldots,y_s) = K(c_1,\ldots,c_t)$, $M = K(z_1, \ldots,z_t) = K(b_1,\ldots,b_s)$. If $F(X_1,\ldots,X_s)$ is an element of the relation ideal for $y_1, \ldots, y_s$ over K, then $F(f_1,\ldots,f_s) = 0$ and $F(b_1,\ldots,b_s) = 0$. This means that there is a natural surjection of $K[y_1,\ldots,y_s]$ to $K[b_1,\ldots,b_s]$. Since trans.deg L = trans.deg $L(x)$ - 1 = trans.deg M, the surjection must be an isomorphism. QED

EXERCISE 4.1

1. Let $K \subseteq L \subseteq M$ be fields. Prove that if $\text{trans.deg}_K M$ is finite or if $\text{trans.deg}_K L$ and $\text{trans.deg}_L M$ are finite, then all three are finite and $\text{trans.deg}_K M = \text{trans.deg}_K L + \text{trans.deg}_L M$.
2. Let T be an extension field of a field K and let L, M be intermediate fields. Prove that $\text{trans.deg}_K L \vee M \leq \text{trans.deg}_K L + \text{trans.deg}_K M$.
3. Give a counterexample to Lemma 4.1.6 in case we drop the condition that $\text{trans.deg}_K L$ is finite.

EXERCISE 4.2

1. Let M, N, M', N' be modules over a field K, and let $f : M \to M'$, $g : N \to N'$ be homomorphisms of K-modules. Then the <u>tensor product</u> $f \otimes g$ of f and g is defined by $(f \otimes g)(\sum m_i \otimes n_i) = \sum (fm_i) \otimes (gn_i)$. Prove that (i) $f \otimes g$ is a K-homomorphism and (ii) if f', g' are K-homomorphisms of M', N', respectively, to some K-modules, then $(f' \otimes g')(f \otimes g) = (f'f) \otimes (g'g)$.
2. Let $M = K[a_1,\ldots,a_n]$, let N be a ring containing a field K, and let I be the relation ideal for $(a_1,\ldots,a_n)$ over K. Prove that $M \otimes_K N \cong N[X_1,\ldots,X_n]/IN[X_1,\ldots,X_n]$. Prove that if furthermore $N = K[Y_1,\ldots,Y_m]/J$, then $M \otimes N \cong K[X_1,\ldots,X_m,Y_1,\ldots,Y_n]/A$ with A generated by I and J.
3. Let T be an extension field of a field K and let L, M be intermediate fields. Consider the conditions:

(i) L and M are linearly disjoint over K.

(ii) If $a_1, \ldots, a_s$ in L are linearly independent over K, then they are linearly independent over M.

(iii) If $b_1, \ldots, b_t$ in M are linearly independent over K, then they are linearly independent over L.

(iv) $L \cap M = K$.

(v) $L \cap M = K$ and M is a Galois extension of K.

Prove (1) that (i), (ii), (iii) are equivalent to each other, (2) that (i) implies (iv), and (3) that (v) implies (i).

4. Let L be an extension field of a field K and let M, N be K-modules. Prove that $(M \otimes_K N) \otimes_K L \cong (M \otimes_K L) \otimes_L (N \otimes_K L)$.

EXERCISE 4.3

1. Assume that L is a finitely generated field over a field K of characteristic $p \neq 0$. Prove that a p-base of L over K consists of a finite number of elements.
2. Prove that if a derivation D of a field K is extended to a derivation of an extension field L, then D can be extended to a derivation of an arbitrary intermediate field.
3. Let R be a ring, and let M be an R-module. A mapping D of R into M is called a derivation of R (with values) in M if it satisfies conditions (i) $D(f + g) = Df + Dg$ and (ii) $D(fg) = f(Dg) + g(Df)$ (for arbitrary $f,g \in R$). Try to adapt the results in Sec. 4.3 to this case.

EXERCISE 4.4

1. Let L be an extension field of a field K. Prove the equivalence of: (i) L is separable over K, (ii) $a_1, \ldots, a_n \in L$ implies that $K(a_1,\ldots,a_n)$ is separable over K, and (iii) $L \otimes_K K^{p^{-\infty}}$ is an integral domain.
2. Let L, M be fields separable over a field K. Prove that if L, M are free over K, then $L \vee M$ is separable over K.
3. Assume that a field L is finitely generated over a field K, and let $z_1, \ldots, z_t$ be a transcendence base of L over K.

Set $M = L \cap K^{p^{-\infty}}$. Prove that $[M : K] \leq i(L/K) \leq [L : K(z_1,\ldots,z_t)]_i$.

EXERCISE 4.5

1. Let K be an arbitrary field of characteristic $p \neq 0$ and let t, u, x, y be algebraically independent elements over K. Let z be a root of $X^p + tx^p + uy^p$ over $M = K(t,u,x,y)$ and set $K^* = K(t,u)$, $L = M(z)$. Prove that K^* is algebraically closed in L, but L is not a regular extension of K^*.
2. Prove that an extension field L of K is a regular extension if and only if $L \otimes_K M$ is an integral domain for every finite algebraic extension M such that M is either separable or purely inseparable.
3. Assume that a field L is finitely generated over a field K. Prove that there is a finite algebraic extension M of K such that (i) the separable closure of K in M is contained in L and (ii) $L \vee M$ is a regular extension of M.

EXERCISE 4.6

1. Let M be a module over a ring R, and let N be a submodule. Prove that M is noetherian if and only if both M/N and N are noetherian.
2. Show that the following statement (*) is false, and that (*) is true under an additional condition that I is nilpotent.

   (*) If an ideal I of a ring R is finitely generated and if R/I is noetherian, then R is noetherian.
3. In the polynomial ring $F = K[X,Y]$ over a field K, show that the ideal $I = X^2F + XYF$ is not primary and that $\sqrt{I}$ is prime.
4. Prove that if M is a maximal ideal of a ring R and if an ideal I is such that $M^n \subseteq I \subseteq M$ (for a natural number n), then I is primary.
5. Prove that if Q is a primary ideal in a ring R and if $a \notin Q$ ($a \in R$), then $Q : a$ is a primary ideal belonging to $\sqrt{Q}$.

6. Let I be an ideal of a noetherian ring R. Prove that, for a prime ideal P of R, P is a prime divisor of I if and only if there is an element a of R such that $I : a = P$.
7. Assume that a ring R is the direct sum of submodules R_i ($i = 0, 1, \ldots, n, \ldots$) such that $R_iR_j \subseteq R_{i+j}$. In this case we say that R is a graded ring and each element of R_n is called a homogeneous element of degree n. An ideal I of R is called homogeneous if I is generated by homogeneous elements. Prove the followings:
   - (i) R_0 is a subring of R having common identity with R.
   - (ii) $M = \sum_{i>0} R_i$ is an ideal of R. If M is generated by homogeneous elements $f_1, \ldots, f_N, \ldots$, then R is generated, as a ring, by $f_1, \ldots, f_N, \ldots$ over R_0.
   - (iii) R is noetherian if and only if R_0 is a noetherian ring and M is finitely generated (as an ideal of R).
   - (iv) Let P, Q be homogeneous ideals of R. P is a prime ideal and Q is P-primary if and only if P is generated by homogeneous elements belonging to $\sqrt{Q}$, and it holds that $fg \in Q$ (f, g homogeneous), $f \notin P$ imply $g \in Q$.
   - (v) If R is noetherian, then every homogeneous ideal I is the intersection of a finite number of homogeneous primary ideals and every prime divisor of I is homogeneous.
8. Let R be a noetherian integral domain. If $a \in R$, $aR \neq R$ and if aR contains a prime ideal P, then either $aR = P$ or $P = 0$.

EXERCISE 4.7

1. Let I be an ideal of a ring R, and let S be a multiplicatively closed subset of R. Let f be the natural homomorphisms of R onto R/I. Prove that if fS consists only of units, then $R_S/IR_S \cong R/I$.

EXERCISE 4.8

1. Assume that an integral domain R' is integral over its subring R which is normal. Prove that if I is an ideal of R ($I \neq R$), then $\operatorname{ht} I = \operatorname{ht} IR'$.

EXERCISE 4.9

1. Prove that, in Theorem 4.9.2, if $\#(K)$ is infinite, then (3) may be replace by (3') $Y_{j+i} = \sum_{j=1}^{n} c_{ij}X_j$ $(c_{ij} \in K)$. Then prove a corresponding result to Corollary 4.9.3.
2. Consider a prime ideal P of height h in the polynomial ring $F = K[X_1,\ldots,X_n]$ over a field K. Prove that the maximal ideal PF_P of F_P is generated by h elements. Prove also that if P is a maximal ideal of F, then P is generated by n elements.

EXERCISE 4.10

1. Let R be a noetherian ring, K its total quotient ring, and b an element of K. Prove that b is integral over R in each of the following two cases: (1) There is a nonzero-divisor a of R such that $ab^n \in R$ for every natural number n; (2) there is an ideal I of R containing a nonzero-divisor such that $bI \subseteq I$.
2. Let R and K be as above. Let R' be a ring integral over R and contained in K. Set $C = \{b \in R \mid bR' \subseteq R\}$. This C is called the <u>conductor</u> of R' over R. Prove that (1) C is the largest member in the set of ideals of R which are ideals of R' simultaneously, (2) $R' = R$ if and only if $C = R$, and (3) R' is a finite R-module if and only if C contains a nonzero-divisor.
3. Generalize Corollary 4.10.6 to arbitrary ideals.

EXERCISE 4.11

1. The real number field does not satisfy any C_i condition.

EXERCISE 4.12

1. Assume that x is transcendental over a field K. Prove that (i) if $a,b,c,d \in K$, $ad \neq bc$, then there is an automorphism σ of $K(x)$ over K such that $\sigma x = (cx + d)/(ax + b)$, and conversely, (ii) every element of $\mathrm{Aut}_K K(x)$ is given in this way.

2. Assume that an extension field $K(x,y)$ of a field K, of characteristic different from 2, is such that the relation ideal for (x,y) is generated by $aX^2 + bY^2 + c$ $(a,b,c \in K)$. Prove that there is t such that $K(x,y) = K(t)$ if and only if $aX^2 + bY^2 + cZ^2 = 0$ has a nontrivial solution in K.
3. Let K be an algebraically closed field of characteristic different from 2. Assume that an extension field $L = K(x,y,z)$ is such that the relation ideal for (x,y,z) is generated by $f_1(Z)X^2 + f_2(Z)Y^2 + f_3(Z)$ with $0 \neq f_i(Z) \in K[Z]$. Prove that there is w such that $L = K(z,w)$.
4. Let K be an infinite field, and let L be a field containing K and contained in $K(x_1,\ldots,x_n)$ with algebraically independent elements $x_1, \ldots, x_n$ over K. Prove that if $t = \text{trans.deg}_K L$, then there are elements $y_1, \ldots, y_t$ such that $L \subseteq K(y_1,\ldots,y_t)$.
5. Prove (2) below, admitting (1).
   (1) Assume that x, y are algebraically independent over an algebraically closed field K. If L is a field such that $K \subset L \subseteq K(x,y)$ and $K(x,y)$ is separably algebraic over L, then $L = K(u,v)$ with certain elements u, v.
   (2) (Zariski-Castelnuovo theorem) Let K be an algebraically closed field of characteristic zero. If L is a field such that $K \subset L \subseteq K(x_1,\ldots,x_n)$ with algebraically independent elements $x_1, \ldots, x_n$ and such that $\text{trans.deg}_K L = 2$, then $L = K(u,v)$ with certain elements u, v.

CHAPTER 5

THEORY OF VALUATIONS

In Secs. 5.1 to 5.6, we deal with valuations of fields whose value groups are contained in the real number field. In the other sections, we include additive valuations (generalized valuations in the sense of Krull). In particular, in Sec. 5.11, we prove some fundamental results on Hensel valuation rings.

5.1 MULTIPLICATIVE VALUATIONS

If a mapping v of a field K into the real number field R satisfies the following three conditions, then v is called a <u>multiplicative valuation</u>, or simply a <u>valuation</u>, of K, and $G = \{vx \mid x \in K, x \neq 0\}$ is called the <u>value group</u> of v.

(1) $va \geq 0$ for every a in K; $va = 0$ if and only if $a = 0$.

(2) If $a,b \in K$, then $(va)(vb) = v(ab)$.

(3) There is a positive number C so that $v(1 + a) \leq C$ for all $a \in K$ such that $va \leq 1$.

Note that, under Condition (2), Condition (3) is equivalent to:

(3') If $a,b \in K$, $va \leq vb$, then $v(a + b) \leq C \cdot vb$.

If a valuation v satisfies (3*) as follows, then we say that v satisfies the <u>triangular inequality</u>:

(3*) If $a,b \in K$, then $v(a + b) \leq va + vb$.

Note that (3*) is stronger than (3) (C can be 2 in the case). If v satisfies a stronger condition (3**) as follows, then we say that v is a <u>nonarchimedean</u> valuation; otherwise v is called an <u>archimedean</u> valuation.

(3**) If $a,b \in K$, then $v(a + b) \le \max\{va,vb\}$.

EXAMPLE 1 If f is an injection of a field K into the complex number field $\mathbb{C}$, then $va = |fa|$ (the absolute value of fa) gives an archimedean valuation v of K. (For the converse, see Sec. 5.6.)

EXAMPLE 2 If we define that $v0 = 0$, $va = 1$ for every $a \neq 0$, then we have a nonarchimedean valuation. This is called the trivial valuation.

EXAMPLE 3 Let p be a prime element of a noetherian integral domain R. Let K be the field of fractions of R. Each element x ($\neq 0$) of K is expressed in the form $p^n a/b$ (n is a rational integer; $a,b \in R$, $ab \notin pR$). Then we define vx to be r^n with a fixed positive real number r which is less than 1. Then v defines a nonarchimedean valuation of K.

This v is called a p-adic valuation of K.

PROOF: (i) Uniqueness of n: Assume that $p^n a/b = p^m c/d$, $m > n$. Then $ad = p^{m-n}bc \in pR$, contradicting our way of expression. (ii) $v0$ must be defined to be 0. Then Conditions (1) and (2) are satisfied. If $vx = r^n \le vy = r^m$, then $n \ge m$. $x = p^n a/b$, $y = p^m c/d$, and $x + y = p^m(p^{n-m}ad + bc)/bd = p^s a'/b'$ with $s \ge m$. Thus $v(x + y) = r^s \le vy$. QED

Let v, w be multiplicative valuations of a field K. We say that v is equivalent to w if there is a positive number s such that $wx = (vx)^s$ for all $x \in K$. Note that, in Example 3 above, valuations equivalent to v are those obtained by changing r ($0 < r < 1$). Note also:

LEMMA 5.1.1 Let v be a multiplicative valuation of a field K, and let s be a positive real number. Define w by $wx = (vx)^s$ for all $x \in K$. Then w is a valuation of K.

This w is denoted by v^s.

PROOF: (1), (2) are obviously maintained, while C in (3) is changed to C^s. QED

LEMMA 5.1.2 Assume that v is a multiplicative valuation of a field K. Then $v1 = 1$, $v(-a) = va$, and $v(a^{-1}) = (va)^{-1}$ for every a ($\neq 0$) in K. If v is nonarchimedean, then $va < vb$ ($a,b \in K$) implies $v(a + b) = vb$.

PROOF: $v1 = v(1^2) = (v1)^2 \neq 0$ and $v1 = 1$. $v(-1)^2 = v1 = 1$ and $v(-1) = 1$; hence $v(-a) = va$. $1 = v(a \cdot a^{-1}) = (va)(v(a^{-1}))$ and $v(a^{-1}) = 1/va$. The last assertion: $vb = v(a + b - a) \leq \max\{v(a + b), v(-a)\}$, which implies $vb \leq v(a + b)$, and we have $vb = v(a + b)$. QED

THEOREM 5.1.3 Let v, w be multiplicative valuations of a field K, and assume that v is not trivial. Then w is equivalent to v if (and only if) $x \in K$, $vx < 1$ imply $wx < 1$.

PROOF: Fix a ($\neq 0$) such that $va < 1$, and set $b = va$, $c = wa$, $s = \log_c b$. We may consider w^s instead of w. Thus we may assume that $va = b = wa$ (<1) and it suffices to show that $v = w$. Let x be an arbitrary element ($\neq 0$) of K. (i) Assume first that $vx < 1$. Set $I = \{(m,n) \mid m,n \in \mathbb{N},\ (va)^{n+1} < (vx)^m \leq (va)^n\}$. Then $(va)^{(n+1)/m} < vx \leq (va)^{n/m}$ and $vx = \lim_{m\to\infty} b^{n/m}$. Our assumption implies that $w(a^{n+1}) < w(x^m) < w(a^{n-1})$, and we have $wx = \lim b^{n/m} = vx$. (ii) If $vx > 1$, then $v(x^{-1}) < 1$, and we have $w(x^{-1}) = v(x^{-1})$; therefore $wx = vx$. (iii) If $vx = 1$, then $v(ax) < 1$ and $v(ax) = w(ax)$. Thus $vx = wx$ in any case and $v = w$. QED

From now on in this section, v is a valuation of a field K, and C is the C in the condition (3).

LEMMA 5.1.4 If $a_1,\ldots,a_n \in K$, $va_i \leq r \in \mathbb{R}$, and $n \leq 2^s$ ($s+1 \in \mathbb{N}$), then $v(a_1 + \cdots + a_n) \leq C^s r$.

PROOF: This is obvious if $s = 0$, and we assume $s > 0$. Set $m = 2^{s-1}$. Then, by induction on s, we have $v(a_1 + \cdots + a_m) \leq C^{s-1}r$, $v(a_{m+1} + \cdots + a_n) \leq C^{s-1}r$. Thus $v(a_1 + \cdots + a_n) \leq C \cdot C^{s-1}r$. QED

LEMMA 5.1.5 Let a_i and r be as above. If $C \leq 2$, then $v(a_1 + \cdots + a_n) \leq nr$, in particular, $v(n \cdot 1) \leq n$.

PROOF: For each natural number t, let $s(t)$ be the natural number such that $2^{s(t)-1} < n^t \leq 2^{s(t)}$. We apply Lemma 5.1.4 to $(a_1 + \cdots + a_n)^t$ which is the sum of n^t terms of the form $a_1^{e_1} \cdots a_n^{e_n}$ $(\sum e_i = t)$ and we have $v(a_1 + \cdots + a_n)^t \leq 2^{s(t)} r^t$. Thus $v(a_1 + \cdots + a_n) \leq 2^{s(t)/t} r$ and therefore $v(a_1 + \cdots + a_n) \leq r \cdot \lim 2^{s(t)/t} = r \cdot 2^{\log_2 n} = rn$. QED

THEOREM 5.1.6 If $C \leq 2$, then v satisfies the triangular inequality.

PROOF: Let $a, b \in K$ and $va = r$, $vb = s$. $(a+b)^n = \sum_t \binom{n}{t} a^{n-t} b^t$. By Lemma 5.1.5, $v\binom{n}{t} a^{n-t} b^t \leq \sum \binom{n}{t} r^{n-t} s^t$. Again by Lemma 5.1.5, we have $v(a+b)^n = v(a+b)^n \leq (n+1) \cdot \max_t \{\binom{n}{t} r^{n-t} s^t\} \leq (n+1) \sum \binom{n}{t} r^{n-t} s^t = (n+1)(r+s)^n$. Thus $v(a+b) \leq (n+1)^{1/n}(r+s)$. Since $\lim (n+1)^{1/n} = 1$, we have $v(a+b) \leq r + s$. QED

COROLLARY 5.1.7 For every valuation v of K, there is a valuation w which is equivalent to v and which statisfies the triangular inequality.

PROOF: Take a positive number t such that $C^t \leq 2$, and set $w = v^t$. QED

THEOREM 5.1.8 Each of the following three conditions characterizes that v is nonarchimedean.†

(1) C can be chosen to be 1.

(2) $v(n \cdot 1) \leq 1$ for any $n \in \mathbb{N}$.

(3) The restriction of v to the prime field P is nonarchimedean.

PROOF: (1) is obvious. Assume that (2) holds good. Considering an equivalent valuation, we may assume that $C \leq 2$. Then we apply our proof of Theorem 5.1.6, assuming that $va = r \geq s = vb$; this time, $v\binom{n}{t} \leq 1$ and $v(a + b)^n \leq (n + 1) \cdot \max\{r^{n-t} s^t\} = (n + 1) r^n$. Thus $v(a + b) \leq r(n + 1)^{1/n}$ and $v(a + b) \leq r$. The converse is obvious, and (2) is proved. If $n \in \mathbb{N}$, then $n \cdot 1 \in P$, and (3) is proved. QED

COROLLARY 5.1.9 If a field has an archimedean valuation, then the characteristic of the field is zero.

---

† Cf. Theorem 5.2.1 and Corollary 5.2.3.

PROOF: Let x be an element of the prime field P. If $\#(P) = p < \infty$, then x^{p-1} is either 1 or 0 and $vx^{p-1} \leq 1$. QED

5.2 VALUATIONS OF THE RATIONAL NUMBER FIELD

In Theorem 5.1.8 and its corollary, we learned that a valuation v of a field K is archimedean if and only if its restriction on the prime field is archimedean; if it is so, then K is of characteristic zero. Therefore it is important to classify valuations of the rational number field $\mathbb{Q}$, and we prove the following theorem.

THEOREM 5.2.1 Let v be a valuation of $\mathbb{Q}$. (1) If v is archimedean, then v is equivalent to the absolute value (i.e., there is a positive real number s such that $vx = |x|^s$). (2) If v is a nontrivial nonarchimedean valuation, then v is a p-adic valuation with a prime number p (cf. Example 3 in Sec. 5.1; the integral domain is the ring $\mathbb{Z}$ of rational integers).

Before proving this theorem, we define the notion of valuation rings:

LEMMA 5.2.2 If v is a nonarchimedean valuation of a field K, then $R_v = \{x \in K \mid vx \leq 1\}$ forms a ring with unique maximal ideal $M_v = \{x \in K \mid vx < 1\}$.

This R_v is called the <u>valuation ring</u> of v, and M_v the <u>valuation ideal</u>.

PROOF: It is easy to see that R_v is a ring and M_v is its ideal in view of Conditions (1), (2), (3**). If $vx = 1$ $(x \in K)$, then $v(x^{-1}) = 1$ and x is a unit in R_v. Thus M_v is the set of nonunits in R_v, and therefore M_v is the unique maximal ideal. QED

PROOF of THEOREM 5.2.1: (2): Assume that v is a nontrivial nonarchimedean valuation of $\mathbb{Q}$ and let R_v, M_v be as above. $\mathbb{Z} \subseteq R_v$, and therefore $P = M_v \cap \mathbb{Z}$ is a prime ideal of $\mathbb{Z}$. Then R_v contains $\mathbb{Z}_P$. Since v is nontrivial, $\mathbb{Z}_P \neq \mathbb{Q}$, and therefore $P = p\mathbb{Z}$ with a prime number p. If $a \in \mathbb{Z}$, $a \notin P$, then $va = 1$; therefore $v(p^n a/b) = vp^n$ $(a,b \in \mathbb{Z};\ a,b \notin P)$, which proves (2).

(1): Assume that v is an archimedean valuation of $\mathbb{Q}$. By Lemma 5.1.7, we may assume that v satisfies the triangular inequality.

Take a natural number $n > 1$, and set $r = \max\{1, vn\}$. Consider n-adic expansion of an arbitrary natural number m, namely, $m = m_t n^t + m_{t-1} n^{t-1} + \cdots + m_1 n + m_0$ $(m_i \in Z;\ 0 \le m_i < n;\ m_t \ne 0)$. (Note the uniquness of this expression.) Then by Lemma 5.1.5, we have

$$vm \le (t+1)\cdot\max\{v(m_i n^i)\} \le (t+1)(n-1)r^t \qquad (n^t \le m < n^{t+1})$$

Applying the same to powers m^u of m, we have $vm^u \le (t(u)+1)(n-1) \times r^{t(u)}$ $(n^{t(u)} \le m^u < n^{t(u)+1})$ and $vm \le (t(u)+1)^{1/u}(n-1)^{1/u} r^{t(u)/u}$. Since $\lim t(u)/u = \log_n m$, we have $vm \le r^{\log_n m}$. If $vn \le 1$, then $r = 1$, and we see that $vm \le 1$ for arbitrary m, which contradicts our assumption that v is archimedean. Thus $vn = r > 1$ for the arbitrary natural number $n > 1$. If we interchange m and n with each other, then we have $r = vn \le s^{\log_m n}$ with $s = vm$. This and the previous inequality $s = vm \le r^{\log_n m}$ imply $s = r^{\log_n m}$. Thus, by $t = \log_r n$, we have $v^t = |\ |$. QED

COROLLARY 5.2.3 A multiplicative valuation v of a field K is nonarchimedean if and only if there is a natural number $n > 1$ such that $v(n\cdot 1) \le 1$.

5.3 TOPOLOGY

We define a topology on a set S by giving a set Ω of open sets whose axiom is as follows: (1) $S \in \Omega$; (2) the empty set is in Ω; (3) if $U, V \in \Omega$, then $U \cap V \in \Omega$; and (4) if $U_\lambda \in \Omega$ $(\lambda \in \Lambda)$, then $\cup_{\lambda\in\Lambda} U_\lambda \in \Omega$.

A set S, on which a topology is defined, is called a topological space; its element is called a point. The complement of an open set is called a closed set. A topology may be defined by giving the set Ω^* of closed sets whose axiom is as follows: (1*) The empty set is in Ω^*; (2*) $S \in \Omega^*$; (3*) if $F, G \in \Omega^*$, then $F \cup G \in \Omega^*$; and (4*) if $F_\lambda \in \Omega^*$ $(\lambda \in \Lambda)$, then $\cap_\lambda F_\lambda \in \Omega^*$.

In a topological space S, an open set U containing a subset M is called an open neighborhood, or simply a neighborhood, of M.†

† In some of the literature, a neighborhood of a point P is a subset containing an open set containing P. But we shall mean by a neighborhood an open neighborhood only.

If we denote by N_P the set of neighborhoods of a point P, then the family of these N_P is called the <u>system of neighborhoods</u> and it enjoys the following five properties as is easily seen: (i) $S \in N_P$; (ii) $U \in N_P$ implies $P \in U$; (iii) $U, V \in N_P$ implies $U \cap V \in N_P$; (iv) if $U_\lambda \in N_P$ $(\lambda \in \Lambda)$, then $\cup_\lambda U_\lambda \in N_P$; and (v) $U \in N_P$, $Q \in U$ imply $U \in N_Q$.

THEOREM 5.3.1 These five conditions characterize the system of neighborhoods. Namely, if a set N_P of subsets of a set S is given to each $P \in S$ and if the set of these N_P enjoys the five properties, then S has a unique topology such that $\{N_P\}$ is the system of neighborhoods.

PROOF: Set $\Omega = \cup_P N_P \cup \{\text{the empty set}\}$. Then Ω satisfies the axiom for the set of open sets. Indeed, (1) and (2) are obvious; if $U, V \in \Omega$ and if $P \in U \cap V$, then $U, V \in N_P$ and $U \cap V \in N_P \subseteq \Omega$; (4) follows from (iv). Thus S has a topology with Ω as the set of open sets. Then (ii) and (v) show that each N_P is the set of neighborhoods of P. The uniqueness of topology is obvious. QED

A set Ω_0 of subsets of a topological space S is called a <u>subbasis</u> of the set Ω of open sets if Ω is the smallest among sets of subsets containing Ω_0 and satisfying the axiom for the set of open sets. Then one sees easily that if a set Ω_0 of subsets of a set S is given, then a topology on S having Ω_0 as a subbasis of the set of open sets is uniquely determined.

A topological space S is called a <u>T_i-space</u> $(i = 0, 1, 2, 3, 4)$ if S satisfies the following condition T_i.

- T_0: If $P, Q \in S$ and if $P \neq Q$, then there is a neighborhood of either P or Q which does not contain the other.
- T_1: If $P, Q \in S$ $(P \neq Q)$, then there is a neighborhood of P which does not contain Q.
- T_2: If $P, Q \in S$ $(P \neq Q)$, then there are neighborhoods U, V of P, Q, respectively, such that $U \cap V$ is empty.
- T_3: If a closed subset F of S does not contain a point P, then there are neighborhoods U, V of P, F such that $U \cap V$ is empty.

T_4: If F, F' are closed subsets of S such that $F \cap F'$ is empty, then there are neighborhoods U, V of F, F' such that $U \cap V$ is empty.

A T_2-space may be called a Hausdorff space. A T_3-space satisfying T_1 is called a regular space. A T_4-space satisfying T_1 is called a normal space. Each of these conditions T_i, regularity, normality is called a separation axiom Though there are many more separation axioms, we do not define them here.

THEOREM 5.3.2 A topological space S is a T_1-space if and only if each point of S forms a closed set.

PROOF: If part: If $P,Q \in S$ $(P \neq Q)$, $S - Q$ is an open set containing P but not Q. Only if part: For each $Q \neq P$ $(P,Q \in S)$, let U_Q be a neighborhood of Q which does not contain P. Then the union of these U_Q is an open set and coincides with $S - P$, and therefore P forms a closed set. QED

Assume that S is a topological space whose topology is given via a function d of $S \times S$ into the real number field R so that:

(1) $d(P,Q) = d(Q,P) \geq 0$ for $P,Q \in S$; $d(P,Q) = 0$ if and only if $P = Q$.

(2) (Triangular inequality) $d(P,Q) + d(Q,R) \geq d(P,R)$ for $P, Q,R \in S$.

(3) For each positive real number r, set $U_r(P) = \{Q \in S \mid d(P, Q) < r\}$. Then a subset U of S is an open set if and only if $P \in U$ implies the existence of r (> 0) such that $U_r(P) \subseteq U$.

In this case, we say that S is a metric space with distance function d. $U_r(P)$ is called the r-neighborhood of P.

Let X be a subset of a topological space S. Then X has a topology such that a subset U is open in X if and only if $U = V \cap X$ with an open subset V in S. This topology on X is called the topology induced from S and X with this topology is called a subspace of S.

THEOREM 5.3.3 Let X be a subspace of a topological space S. Then (1) a subset F of X is closed in X if and only if $X \cap F' = F$ with a closed set F' in S; (2) if S is a metric space, then so is X; (3) if S is a T_i-space with $i \leq 3$, then so is X; and (4) the following implications hold: Metric space $\Rightarrow$ normal space $\Rightarrow$ regular space $\Rightarrow$ Hausdorff space $\Rightarrow$ T_1-space $\Rightarrow$ T_0-Space.

PROOF: (1): $U = X - F$ is an open set in X; hence $U = U' \cap X$ with an open set U' in S and $F = (S - U') \cap X$. (2): Obvious. (3): Obvious for $i \leq 2$; in the case where $i = 3$, we see the result by (1) above. (4): In view of Theorem 5.3.2, we see the implications except for the first. Assume that S is a metric space with distance function d. For $P \in S$ and $F \subseteq S$, we define the <u>distance</u> $d(P,F)$ to be the infimum of $\{d(P,Q) \mid Q \in F\}$. Then:

LEMMA 5.3.4 If F is a closed set and if $P \notin F$, then $d(P,F) > 0$.

PROOF: $S - F$ is an open set containing P, and hence $U_r(P) \subseteq S - F$ for some $r > 0$. This means that $d(P,Q) \geq r$ for any $Q \in F$. QED

Now we prove that every metric space is a normal space. Let F and F' be closed sets such that $F \cap F'$ is empty. Take the $d(P,F')/2$-neighborhood of P for each $P \in F$, and let U be the union of all of them. Similarly, take the $d(Q,F)/2$-neighborhood of Q for each $Q \in F'$. and let U' be the union of all of them. Then U, U' are neighborhoods of F, F', and $U \cap U'$ is empty. QED

A mapping f of a topological space S to a topological space T is said to be <u>continuous</u> if for each neighborhood V of fP ($P \in S$), there is a neighborhood U of P such that $fU \subseteq V$.

THEOREM 5.3.5 Let f be a mapping of a topological space S into a topological space T. Then f is continuous if and only if $f^{-1}(V) = \{P \in S \mid fP \in V\}$ is an open set in S for every open set V in T.

PROOF: If part: $U = f^{-1}(V)$ satisfies the requirement. Only if part: Let $P \in f^{-1}(V)$. Then V is a neighborhood of fP, and there is a neighborhood U_P of P such that $fU_P \subseteq V$. Then $U_P \subseteq f^{-1}(V)$; therefore $f^{-1}(V)$ is the union of all U_P and hence is open. QED

A mapping f of a topological space S to a topological space T is called a homeomorphism if (i) f gives a one-one correspondence between S and T and (ii) both f and f^{-1} are continuous. If such f exists, then we say that S is homeomorphic to T. In this case, a subset U of S is an open set if and only if fU is open in T.

Consider a family of topological spaces S_λ $(\lambda \in \Lambda)$. We introduce a topology on the product set $S^* = \Pi S_\lambda$ by taking $\Omega_0 = \{p_\lambda^{-1}(U_\lambda) \mid \lambda \in \Lambda,\ U_\lambda$ an open set in $S_\lambda\}$ as a subbasis for the set of open sets, where p_λ denotes the projection of S^* to S_λ. This S^* is called the product space of these S_λ and is denoted by the same symbol $\Pi_\lambda S_\lambda$; in the finite number case, we may use notation such as $S_1 \times \cdots \times S_n$.

THEOREM 5.3.6 Let $S_1, \ldots, S_n$ be topological spaces. A subset U of the product space $S^* = S_1 \times \cdots \times S_n$ is an open set if and only if it is the union of certain sets of the form $U_1 \times \cdots \times U_n$ with open sets U_i in S_i.

PROOF: Since $U_1 \times \cdots \times U_n = \cap_i (S_1 \times \cdots \times S_{i-1} \times U_i \times S_{i+1} \times \cdots \times S_n)$, we see that each $U_1 \times \cdots \times U_n$ is an open set in S^* and the if part follows from this. Conversely, let Ω' be the set of U obtained in the same manner as in the theorem. Ω' contains the subbasis Ω_0 in the definition. If U, V are in Ω', then U, V are the unions of $U_{1\lambda} \times \cdots \times U_{n\lambda}$ $(\lambda \in \Lambda)$ and $V_{1\mu} \times \cdots \times V_{n\mu}$ $(\mu \in \Lambda')$, respectively. Then $U \cap V = \cup_{\lambda\mu} ((U_{1\lambda} \times \cdots \times U_{n\lambda}) \cap (V_{1\mu} \times \cdots \times U_{n\mu})) = \cup_{\lambda\mu} (\cap_i ((S_1 \times \cdots \times S_{i-1} \times U_{i\lambda} \times S_{i+1} \times \cdots \times S_n) \cap (S_1 \times \cdots \times S_{i-1} \times V_{i\mu} \times S_{i+1} \times \cdots \times S_n)) = \cup(\cap_i (S_1 \times \cdots \times S_{i-1} \times (U_{i\lambda} \cap V_{i\mu}) \times S_{i+1} \times \cdots \times S_n)) = \cup_{\lambda\mu} (U_{1\lambda} \cap V_{1\mu}) \times \cdots \times (U_{n\lambda} \cap V_{n\mu}) \in \Omega'$. Thus Ω' satisfies the axiom for the set of open sets, and the converse is proved. QED

Let X be a subset of a topological space S. The intersection X^* of all closed sets containing X is the smallest closed set containing X, and X^* is called the closure of X. We say that X is dense in S if $X^* = S$. Under the notation, we have the following:

THEOREM 5.3.7 $P \in X^*$ if and only if every neighborhood of P meets X.

PROOF: $Q \notin X^*$ if and only if there is a closed set F such that $Q \notin F \supseteq X$, or equivalently, there is a neighborhood U of Q which does not meet X. QED

Assume that two topologies, say τ_1 and τ_2, are given on a set S. If the identity mapping f $(fx = x)$ on S is continuous as (S with τ_1) $\to$ (S with τ_2), then we say that τ_1 is stronger than τ_2 or τ_2 is weaker than τ_1. (It would be more rigorous to say "stronger or equal" or "weaker or equal," respectively.) The strongest topology is the case where every subset is an open set; in this case the topology is said to be discrete. The weakest topology is the case where S and the empty set are the only open sets; this case is not important because S is not a T_0-space if $\#(S) \geq 2$.

THEOREM 5.3.8 For a topological space S, the following two conditions are equivalent to each other; if S satisfies them, then we say that S is compact:†

(1) If U_λ $(\lambda \in \Lambda)$ are open sets in S and if $S = \cup U_\lambda$, then there are a finite number of members $\lambda_1, \ldots, \lambda_n$ of Λ such that $S = \cup_i U_{\lambda_i}$.

(2) If F_λ $(\lambda \in \Lambda)$ are closed sets in S satisfying finite intersection property (i.e., any finite number of members have a commonpoint), then all F_λ have a common point.

PROOF: Assume that (1) holds. Set $V_\lambda = S - F_\lambda$. If $\cup V_\lambda = S$, then (1) shows that $S = \cup_{i=1}^n V_{\lambda_i}$ with $\lambda_i \in \Lambda$, which means that $\cap_i F_{\lambda_i}$ is empty. This contradicts our assumption; $\cup V_\lambda \neq S$ and $\cap F_\lambda$ is not empty. The converse is proved similarly. QED

One familiar example of a compact space is a bounded closed set in a euclidean space; in that case, (1) corresponds to the usual form

† In some of the literature, a compact space means a compact T_2-space.

of the covering theorem of Heine-Borel. Note also that S with discrete topology is compact if and only if $\#(S)$ is finite.

5.4 TOPOLOGICAL GROUPS AND TOPOLOGICAL FIELDS

Let G be a semigroup on which a topology is defined. We say that the operation, say multiplication, on G is continuous if the mapping f of $G \times G$ (as the product space) to G such that $f(a,b) = ab$ is continuous. This condition is equivalent to: If $a,b \in G$ and if a neighborhood U of ab is given, then there are neighborhoods V, W of a, b, respectively such that $VW = \{vw \mid v \in V, w \in W\} \subseteq U$.

We say that a group G on which a topology is defined is a topological group if (i) G is a T_0-space, (ii) the operation, say multiplication, on G is continuous, and (iii) the mapping t of G to G which maps each x to x^{-1} (in case of addition, x to $-x$) is continuous.

THEOREM 5.4.1 Let G be a group on which a topology is defined.

(1) If the condition (ii) above is satisfied, then, for each $g \in G$, the mappings $r_g : x \rightsquigarrow xg$ and $s_g : x \rightsquigarrow gx$ are homeomorphisms of G to G.

(2) If the condition (iii) above is satisfied, then the mapping t is a homeomorphism of G to G.

PROOF: (1): Obviously r_g and s_g are continuous for every g; the inverses are $r_{g^{-1}}$ and $s_{g^{-1}}$ which are continuous because $g^{-1} \in G$.

(2): t is continuous, and t^2 is the identity mapping. QED

REMARK: The converse of (1) above is false. For instance, let G be the additive group of rational numbers on which topology is defined so that a subset U is open if and only if either U is empty or $\#(G - U)$ is finite. Then G satisfies (i), (iii) in the definition of a topological group, and r_g, s_g are continuous for every $g \in G$. But, if U, V are neighborhoods of 0, then $U + V = G$, and therefore the condition (ii) is not satisfied.

COROLLARY 5.4.2 Let G be a topological group.

(1) Let N_x be the set of neighborhoods of x for each $x \in G$. Then $N_x = x(N_1) = (N_1)x$; for a subset U of G, $U \in N_x$ if and only if $U^{-1} = \{x^{-1} \mid x \in U\}$ is in $N_{x^{-1}}$.

(2) G is a T_2-space.†

PROOF: (1) is obvious. As for (2), let $x, y \in G$, $x \neq y$. By T_0, there is a $U \in N_1$ such that either $x \notin yU$ or $y \notin xU$. We assume that $y \notin xU$. By the continuity of the multiplication, there are $V, W \in N_1$ such that $VW \subseteq U$. By (1), $W^{-1} \in N_1$ and xV, yW^{-1} are neighborhoods of x, y, respectively. If $z \in xV \cap yW^{-1}$, then $z = xv = yw^{-1}$ ($v \in V$, $w \in W$) and $y = xvw \in xVW \subseteq xU$; this is a contradiction, and $xV \cap yW^{-1}$ is empty. QED

Let G be a topological group, and let N^* be a set consisting only of neighborhoods of 1. If every neighborhood of 1 contains some member of N^*, then N^* is called a fundamental system of neighborhoods of the identity.

THEOREM 5.4.3 If N^* is a fundamental system of neighborhoods of 1 in a topological group G, then N^* satisfies the following conditions (i) to (vi). Conversely, if a set N^* of subsets of a group G satisfies these six conditions, then a unique topology is defined on G so that G is a topological group with N^* as a fundamental system of neighborhoods of 1.

(i) $U \in N^*$ implies $1 \in U$.

(ii) $1 \neq x \in G$ implies the existence of $U \in N^*$ such that $x \notin U$.

(iii) If $U, V \in N^*$, then there is $W \in N^*$ such that $W \subseteq U \cap V$.

(iv) If $U \in N^*$, then there is $V \in N^*$ such that $VV = \{xy \mid x, y \in V\} \subseteq U$.

(v) If $U \in N^*$, then there is $V \in N^*$ such that $V^{-1} = \{x^{-1} \mid x \in V\} \subseteq U$.

(vi) If $a, b \in G$, $U \in N^*$ and if $1 \in aUb$, then there is $V \in N^*$ such that $V \subseteq aUb$.

PROOF: The first half: (i), (iii), and (vi) follow from the

† It is known that a topological group is a completely regular space.

fact that each member is a neighborhood of 1. (ii) follows from T_2. (iv) follows from the continuity of the multiplication, and (v) from that of $x \rightsquigarrow x^{-1}$.

The last half: Let Ω be the set of subsets which are unions of sets of the form aU ($a \in G$, $U \in N^*$), and let Ω' be the set similarly obtained by sets of the form Ua. If $b \in X \in \Omega'$, then $b \in Ua \subseteq X$ ($U \in N^*$, $a \in G$). $1 \in b^{-1}Ua$, and there is $V \in N^*$ such that $V \subseteq b^{-1}Ua$. Then $bV \subseteq Ua$. Thus X is the union of such bV, and $X \in \Omega$. Thus $\Omega' \subseteq \Omega$. Similarly, $\Omega \subseteq \Omega'$ and we have $\Omega = \Omega'$. If $U_1a_1 \cap \cdots \cap U_ma_m$ ($U_i \in N^*$, $a_i \in G$) contains an element b, then $1 \in U_ia_ib^{-1}$, and there are $V_i \in N^*$ such that $V_i \subseteq U_ia_ib^{-1}$ [by (vi)]. Then there is $W \in N^*$ such that $W \subseteq \cap V_i \subseteq \cap U_ia_ib^{-1}$. Thus $Wb \subseteq \cap U_ia_i$, and we see that $\cap U_ia_i \in \Omega$. Using this fact, one sees easily that Ω satisfies the axiom for the set of open sets. Thus we define a topology on G. An arbitrary neighborhood of a ($\in G$) contains aV and Va for some $V \in N^*$. If W' is a neighborhood of ab ($a,b \in G$), then there is $U \in N^*$ such that $Uab \subseteq W'$. Then there are $V \in N^*$ such that $VV \subseteq a^{-1}Ua$ [by (iv) and (vi)] and $V' \in N^*$ such that $V' \subseteq aVa^{-1}$. Then $(V'a)(Vb) \subseteq (aV)(Vb) = aVVb \subseteq Uab$. This shows the continuity of the multiplication. The continuity of $x \rightsquigarrow x^{-1}$ follows from (v). That G is a T_0-space follows from (ii). Thus G is a topological group. N^* is obviously a fundemental system of neighborhoods of 1. The uniqueness of the required topology is also obvious in view of our definition of Ω. QED

If a neighborhood U of 1 is such that $U^{-1} = U$, then we say that U is a symmetric neighborhood of 1. If a fundamental system of neighborhoods of 1 consists only of symmetric neighborhoods, then we call the system a fundamental system of symmetric neighborhoods.

LEMMA 5.4.4 If N^* is a fundamental system of neighborhoods of 1 in a topological group G, then $N' = \{U \cap U^{-1} \mid U \in N^*\}$ is a fundamental system of symmetric neighborhoods.

PROOF: This follows from Condition (v). QED

THEOREM 5.4.5 If a subgroup H of a topological group G is an open set, then H is a closed set.

PROOF: Each Hx $(x \in G,\ x \notin H)$ is an open set, and the union V of them is an open set. Obviously $V = G - H$ and H is a closed set. QED

A ring with topology is called a <u>topological ring</u> if it is a topological group with respect to its addition and if the multiplication is continuous. A field K with topology is called a <u>topological field</u> if both K with addition and $K - \{0\}$ with multiplication are topological groups.

THEOREM 5.4.6 Let v be a multiplicative valuation of a field K. Let w be a valuation of K equivalent to v and satisfying the triangular inequality. Then K becomes a metric space with distance function d such that $d(x,y) = w(x - y)$ and K becomes a topological field.

The topology on K defined above is called the <u>topology defined by</u> v.

PROOF: It is obvious that K is a metric space with distance function d. $w((a + x) + (b + y) - (a + b)) = w(x + y) \leq w(x) + w(y)$ shows the continuity of addition; $w(-x) = wx$ shows the continuity of $x \rightsquigarrow -x$; $w((a + x)(b + y) - ab) = w(ay + bx + xy) \leq (wy)(wa) + (wx)(wb) + (wx)(wy)$ shows the continuity of the multiplication; lastly, $w((a + x)^{-1} - a^{-1}) = w(x/(a + x)a) = (wx)(wa)^{-1}(w(a + x))^{-1}$ shows the continuity of $a \rightsquigarrow a^{-1}$. Thus K is a topological field. QED

REMARK: Note that although the distance function d depends on the choice of w, topology of K does not depend on the choice of w.

THEOREM 5.4.7 Let K and v be as above, and let K' be a subfield of K. Then the restriction v' of v on K' defines a topology as a subspace of K.

PROOF: These topologies on K and K' are defined by the same distance function. QED

5.5 COMPLETIONS

Let S be a metric space with distance function d. A sequence P_1, P_2, ..., P_n, ... of points of S, which is denoted by $\{P_n\}$, is called a <u>Cauchy sequence</u> if, for every positive real number r, there is a natural number N such that $d(P_m,P_n) < r$ for all $m,n > N$. If it holds that $d(P_m,P_n) < 2^{-n}$ for every $m > n$, then we call $\{P_n\}$ a <u>regular</u> Cauchy sequence. We say that a point P is a <u>limiting point</u> of $\{P_n\}$ or that $\{P_n\}$ <u>converges</u> to P if, for every positive number r, there is a natural number N such that $d(P,P_n) < r$ for every $n > N$. In this case, we write $P = \lim P_n$. Note that, in this case, $\{P_n\}$ is a Cauchy sequence, and P is uniquely determind by $\{P_n\}$. S is said to be <u>complete</u> if every Cauchy sequence in S has a limiting point.

We say that a sequence $\{P_n\}$ is <u>equivalent</u> to a sequence $\{Q_n\}$ if the sequence $\{d(P_n,Q_n)\}$ converges to 0. In this case, if one of $\{P_n\}$, $\{Q_n\}$ is a Cauchy sequence, then so is the other.

LEMMA 5.5.1 (1) Let $\{P_n\}$ be a Cauchy sequence which converges to a point P. Then a sequence $\{Q_n\}$ is equivalent to $\{P_n\}$ if and only if $\{Q_n\}$ converges to P. (2) If $\{P_n\}$ is a Cauchy sequence, then there is a regular Cauchy sequence $\{Q_n\}$ which is equivalent to $\{P_n\}$.

PROOF: (1) is obvious, and we observe (2). For each 2^{-n}, let $N(n)$ be a natural number such that $d(P_m,P_s) < 2^{-n}$ for all $m, s > N(n)$. Let $m_1, m_2, \ldots$ be an ascending sequence of natural numbers such that $m_i > N(i)$, and set $Q_n = P_{m_n}$. Then $\{Q_n\}$ is the required sequence. QED

Let v be a multiplicative valuation of a field K. We say that v is <u>complete</u> or that K is <u>complete</u> with respect to v if K is complete under the metric given by Theorem 5.4.6.

THEOREM 5.5.2 Let v be a multiplicative valuation of a field K. Then there is a field K^* with a valuation v^* satisfying the following conditions (1) to (3). Such a pair (K^*,v^*) is unique within K-isomorphisms. (K^*,v^*) is called the <u>completion</u> of (K,v):

(1) v^* is complete, (2) K is a subfield and a dense subset of K^*, and (3) v^* is an <u>extension</u> of v (i.e., the restriction of v^* on K is v).

PROOF: In view of our definition of the topology, we may assume that v satisfies the triangular inequality. Let S be the set of Cauchy sequences in K. We introduce addition and multiplication in S by: $\{a_n\} + \{b_n\} = \{a_n + b_n\}$, $\{a_n\}\{b_n\} = \{a_n b_n\}$. Then S becomes a ring whose identity and zero are $\{1,1\ldots\}$ and $\{0,0,\ldots\}$. The set I of null-sequences (i.e., sequences which converge to 0) forms and ideal of S, and two elements $\{a_n\}$, $\{b_n\}$ are equivalent to each other if and only if $\{a_n\} - \{b_n\} \in I$. Therefore the ring $K^* = S/I$ is the set of equivalence classes of Cauchy sequences in K. The mapping f of K into S such that $fa = \{a,a..\}$ is a homomorphism. Since K is a field, we see that f induces an injection of K into K^*. Therefore we may regard K as a subfield of K^*. We note here that for every $\{a_n\} \in S$, the sequence $\{v(a_n)\}$ is a Cauchy sequence (in the set of real numbers), because $|v(a_n) - v(a_m)| \leq v(a_n - a_m)$ by the triangular inequality.

(i) That K^* is a field: If $0 \neq a^* \in K^*$, then a^* is the class of a sequence $\{a_n\}$ $(\in S, \notin I)$. Then there is a positive number r such that $v(a_n) \geq r$ for large n, because $\{a_n\} \notin I$. Taking an equivalent sequence, we may assume that $v(a_n) \geq r$ for every n. Consider the sequence $\{a_n^{-1}\}$. Since $d(a_m^{-1}, a_n^{-1}) = v((a_m - a_n)/a_m a_n \leq v(a_m - a_n)r^2$, we see that $\{a_n^{-1}\}$ is a Cauchy sequence whose class is the inverse of a^*.

(ii) Valuation v^*: For the class a^* of $\{a_n\} \in S$, we define v^*a^* to be $\lim v(a_n)$. This is well defined because if a^* is the class of $\{b_n\}$, then $\{a_n\} - \{b_n\}$ is a null-sequence and $\lim v(a_n) = \lim v(b_n)$. If c^* is the class of $\{c_n\} \in S$, then $v^*(a^*c^*) = \lim v(a_n c_n) = (\lim v(a_n))(\lim v(c_n)) = (v^*a^*)(v^*c^*)$, $v^*(a^* + c^*) = \lim v(a_n + c_n) \leq \lim (v(a_n) + v(c_n)) = v^*a^* + v^*c^*$. $v^*0 = 0$ obviously. As we saw in (i) above, if $0 \neq a^* \in K^*$, then $v^*a^* > 0$. Thus v^* is a valuation of K^* and is obviously an extension of v.

(iii) Dense property: Let a^* be the class of $\{a_n\} \in S$. Then $v^*(a^* - a_m) = \lim_n v(a_n - a_m)$. For each positive number r, there is a natural number $N(r)$ such that $v(a_n - a_m) < r$ for all m, $n > N(r)$.

Therefore, if $m > N(r)$, then $v^*(a^* - a_m) \leq r$. This shows that $a^* = \lim a_n$ and K is a dense subset of K^*.

(iv) Completeness: Let $\{a^*_n\}$ be a Cauchy sequence in K^*. Let $\{a_{nm} \mid m = 1, 2, \ldots\}$ be a regular Cauchy sequence whose class is a^*_n (Lemma 5.5.1). Consider the sequence $\{a_{nn}\}$ in K. Since $\{a^*_n\}$ is a Cauchy sequence, for each positive number r, there is a natural number $M(r)$ such that $v^*(a^*_m - a^*_n) < r$ for all $m,n > M(r)$, namely, there is a natural number $M' = M'(r,m,n)$ such that $v(a_{mt} - a_{nt}) < r$ for $t > M'$ $(m,n > M(r))$. Since every $\{a_{nm} \mid m = 1, 2, \ldots\}$ is a regular Cauchy sequence, $v(a_{ns} - a_{nt}) < 2^{-t}$ if $s > t$. Therefore, for sufficiently large t, we have $v(a_{mm} - a_{nn}) = v(a_{mm} - a_{mt} + a_{mt} - a_{nt} + a_{nt} - a_{nn}) \leq v(a_{mm} - a_{mt}) + v(a_{mt} - a_{nt}) + v(a_{nt} - a_{nn}) < 2^{-m} + 2^{-n} + v(a_{mt} - a_{nt})$. Thus we see that $\{a_{nn}\}$ is a Cauchy sequence, and hence its class b^* is in K^*. Then $b^* = \lim a_{nn}$ and $v^*(a^*_n - b^*) = \lim_m v(a_{nm} - a_{mm}) = \lim_m v(a_{nm} - a_{nn} + a_{nn} - a_{mm}) < 2^{-n} + \lim_m v(a_{nn} - a_{mm})$. Thus $v^*(a^*_n - b^*)$ tends to 0, and $b^* = \lim a^*_n$.

(v) Uniqueness: Let (K',v') be a completion of (K, v). Note that if $\{a_n\} \in S$, then $v'(\lim a_n) = \lim v(a_n)$. Let g be the mapping of S into K' which maps $\{a_n\}$ to $\lim a_n$ in K'. Then the kernel of g is I. Since K is dense in K', we see that $gS = K'$. Thus $K' \cong K^*$. By the equality that $v'(\lim a_n) = \lim v(a_n)$, we see that v' corresponds to v^* by the isomorphism. QED

THEOREM 5.5.3 Let v be a valuation of a field K, and let w be an extension of v to an extension field L. Let K^* be the closure of K under the topology of L defined by w. If w is complete, then K^*, with the restriction v^* of w, is the completion of K.

PROOF: Let S be the set of Cauchy sequences in K which is a subspace of L (Theorem 5.4.7). Let f be the mapping of S into L such that $f(\{a_n\}) = \lim a_n$. Then $K^* = fS$, and we see that K^* here is isomorphic to the K^* in Theorem 5.5.2, and we can complete the proof easily. QED

There is a big difference between archimedean and nonarchimedean valuations on the matter of value groups. Namely:

THEOREM 5.5.4 Let K, v, K*, v* be as before. (1) If v is nonarchimedean valuation, and if $0 \neq a^* \in K^*$ ($a^* = \lim a_n$), then $v^*(a^*) = v(a_n)$ for sufficiently large n. Therefore the value group of v* coincides with that of v. (2) If v is archimedean, then the value group of v* is the group R^+ of positive real numbers.

PROOF: (1): Let r be a real number such that $\lim v(a_n) > 2r > 0$, and let N be a natural number such that $v(a_n - a_m) < r$ and $v(a_m) > \lim v(a_s) - r$ for $m,n > N$. Then $v(a_m) > r$ and $v(a_n) = v(a_n - a_m + a_m) = v(a_m)$ (Lemma 5.1.2). Thus $\lim v(a_n) = v(a_m)$ for large m.

(2): K contains the rational number field Q, and the restriction v' of v on Q is equivalent to the absolute value (Theorem 5.2.1). Thus v(K) contains a dense subset in R^+, and we see the result. QED

5.6 ARCHIMEDEAN VALUATION AND ABSOLUTE VALUE

We aim to prove the following theorem in this section:

THEOREM 5.6.1 If v is an archimedean valuation of a field K, then there are an injection f of K into the complex number field C and a positive real number t such that $vx = |fx|^t$ ($|\ |$ denotes the absolute value). If furthermore K is complete, then fK is either C or the real number field R.

PROOF: Considering the completion, we may assume that K is complete. K contains the rational number field Q, and the restriction w of v on Q is equivalent to $|\ |$ (Theorem 5.2.1); we may assume that $w = |\ |$. K contains the completion of Q (Theorem 5.5.3), which is R. Thus K contains R, and $vx = |x|$ for every $x \in R$. Then, it suffices to prove the following (1), (2);

(1) If $\sqrt{-1} \in K$ and if $vx = |x|$ for every $x \in C$, then $K = C$.

(2) If $\sqrt{-1} \notin K$ then v is uniquely extended to a valuation of $K(\sqrt{-1})$ and the extension is unique and is complete.

Indeed, (*) if $\sqrt{-1} \in K$, then $|\ |$ in R can be extended to

$|\ |$ in $\mathbb{C}$; hence, by (2), the restriction of v on $\mathbb{C}$ must be $|\ |$ and $K = \mathbb{C}$ by (1). If $\sqrt{-1} \notin K$, then $K(\sqrt{-1}) = \mathbb{C}$ by (*), and $K = \mathbb{R}$ because $K \supseteq \mathbb{R}$.

PROOF of (1): We assume that the following fact is known to the readers: If f is a real-valued continuous function defined on a bounded closed set F on the real plane, then $f(F)$ has the least value.†

Assume that $K \neq \mathbb{C}$, and let a be an element of K which is not in $\mathbb{C}$. Let $f(z) = v(z - a)$ for $z \in \mathbb{C}$. Then f is a continuous function, and $f(z) > 0$ for any $z \in \mathbb{C}$. $f(z) = v(z - a) \geq vz - va = |z| - va$, and therefore if $|z|$ is sufficiently large, then $f(z)$ is large. Let N be a sufficiently large number, and consider f on $F = \{z \in \mathbb{C} \mid |z| \leq N\}$. Then the least value $f(c)$ $(c \in F)$ in $f(F)$ is the least value in $f(\mathbb{C})$. Set $r = f(c)$; we want to show:

(*) If $x,y \in \mathbb{C}$, $f(x) = r$ and $|y| < r$, then $f(x - y) = r$.

Indeed, setting $b = x - a$, we have $vb = r$ by definition. For each natural number n, we consider $b^n - y^n$. Let ζ be a primitive n-th root of unity. Then $b^n - y^n = (b - y)(b - \zeta y)\cdots(b - \zeta^{n-1}y)$ and $v(b^n - y^n) = v(b - y)v(b - \zeta y)\cdots v(b - \zeta^{n-1}y)$. Because of the minimality of r, we have $v(b^n - y^n) \geq v(b - y)r^{n-1}$, and therefore $v(b - (y^n/b^{n-1})) \geq v(b - y)$. If follows that $vb + (vy)^n/(vb)^{n-1} \geq v(b - y) = f(x - y)$. Since $vy < r = vb$, $(vy)^n/(vb)^{n-1}$ tends to 0 if n tends to infinity. Thus $r = vb \geq f(x - y)$. By the minimality of r, we have $f(x - y) = r$. This completes the proof of (*) above.

Now take $z \in \mathbb{C}$ such that $|z|$ is sufficiently large, and take a natural number m such that $|z/m| < r$. By (*) above, $f(c - i(z/m)) = r$ for $i = 0, 1, \ldots$ (by induction on i). By the case $i = m$, we have $f(c - z) = r$. On the other hand, $f(c - z) = v(c - z - a) \geq vz - v(c - a) \geq |z| - |c| - va$, and therefore $r \geq |z| - |c| - va$, contradicting that $|z|$ is large.

PROOF of (2): Assume first the existence of an extension w of

---

† Using Theorem 5.3.5, we can prove that if f is a continuous mapping of a compact space S to a topological space T, then fT is compact. The fact referred to here follows easily from this result.

v. Then, considering the norm $N = N_{K(\sqrt{-1})/K}$, we want to show that $wz = \sqrt{v(N(z))}$. Each element of $K(\sqrt{-1})$ is expressed uniquely in the form $x + y\sqrt{-1}$ $(x,y \in K)$. Then $N(x + y\sqrt{-1}) = x^2 + y^2$ because the conjugates of $x + y\sqrt{-1}$ over K are $x \pm y\sqrt{-1}$. Therefore, if we see that $w(x + y\sqrt{-1}) = w(x - y\sqrt{-1})$, then we have the equality $wz = \sqrt{v(N(z))}$. Assume for a moment that $w(a + b\sqrt{-1}) < w(a - b\sqrt{-1})$ for some $a,b \in K$. Then $z = (a + b\sqrt{-1})/(a - b\sqrt{-1})$ is an element of $K(\sqrt{-1})$ such that $N(z) = 1$ and $wz < 1$. Considering powers of z, we see that there is a sequence $z_1, \ldots, z_n, \ldots$ of elements of $K(\sqrt{-1})$ such that $N(z_n) = 1$ and $w(z_n) < 2^{-n}$. Write $z_n = a_n + b_n\sqrt{-1}$. Considering $z_n\sqrt{-1}$ if necessary, we may assume that $v(a_n) \geq v(b_n)$ for every n. Since $N(z_n) = 1$, we have $2^n < w(a_n - b_n\sqrt{-1}) \leq Cw(a_n)$ [where C is the constant with respect to w in Condition (3) of a valuation]. Therefore $w(a_n)$ $[= v(a_n)]$ must be very large if n is very large. Therefore, considering $y_n = z_n/a_n = 1 + (b_n/a_n)\sqrt{-1}$, we have $w(y_n) < 2^{-n}$. Therefore the sequence $\{y_n\}$ converges to 0, which implies that $\{(b_n/a_n)\sqrt{-1}\}$ converges to -1 and $\{b_n/a_n\}$ converges to $\sqrt{-1}$; this contradicts our assumption that K is complete and $\sqrt{-1} \notin K$.

Now let us show the mapping w such that $wz = \sqrt{v(N(z))}$ is a valuation. For our purpose, we may assume that v satisfies the triangular inequality. Since $N(zz') = N(z)N(z')$ and since $N(z) = 0$ if and only if $z = 0$, it suffices to show the existence of the constant C such that $wz \leq 1$ implies $w(1 + z) \leq C$. Assume that there is no such C. Then for each natural number n, there is $z_n = a_n + b_n\sqrt{-1}$ such that $w(z_n) \leq 1$, $w(1 + z_n) > 2^n$; namely, $v(a_n^2 + b_n^2) \leq 1$, $v((1 + a_n)^2 + b_n^2) > 2^{2n}$. Since v satisfies the triangular inequality, we have $v(1 + 2a_n) > 2^{2n} - 1$, and $v(2a_n) > 2^{2n} - 2$. Then the sequence $\{z_n/a_n\}$ tends to 0, and we have a contradiction as in the first half of the present proof.

The completeness of $K(\sqrt{-1})$ is easy and we leave it to the reader. QED

5.7 ADDITIVE VALUATIONS AND VALUATION RINGS

The additive group of real numbers and the multiplicative group of positive real numbers are groups which are ordered sets whose orders are compatible with group operations. Generalizing the situation, we define the notion of an ordered additive group as follows: It is an additive group G such that (i) G is a linearly ordered set, (ii) for $a,b \in G$, $a > b$ if and only if $-a < -b$, and (iii) $a \geq b$, $c \geq d$ $(a,b,c,d \in G)$ imply $a + c \geq b + d$. (An ordered group is defined similarly.)

v is an additive valuation, often called simply a valuation, of a field K if v is mapping of K into $G \cup \{\infty\}$, where G is an ordered additive group and $\infty > a$ for every $a \in G$, such that (1) $vx = \infty$ if and only if $x = 0$, (2) $v(xy) = vx + vy$ (for $x,y \in K$, $xy \neq 0$; by defining $a + \infty = \infty + a = \infty$ for $a \in G$ and $\infty + \infty = \infty$, we may regard that the equality holds good for arbitrary $x, y \in K$), and (3) $v(x + y) \geq \min \{vx,vy\}$.

In this case, $\{vx \mid 0 \neq x \in K\}$ forms a subgroup of G which is called the value group of v.

An easy example of an additive valuation is obtained from a nonarchimedean valuation. Nemely, if v is a nonarchimedean valuation of a field K, then taking a real number r such that $1 > r > 0$, we set $wx = \log_r vx$. Then this w gives an additive valuation of K whose value group is contained in the additive group $\mathbb{R}$ of real numbers. Conversely, if w is an additive valuation of a field K such that the value group is contained in $\mathbb{R}$, then, by setting $vx = r^{wx}$, we have a multiplicative valuation v of K. Thus additive valuations with value groups within $\mathbb{R}$ are substantially the same as nonarchimedean valuations (cf. Theorem 5.7.5 below). Note here that the difference is not just the expression of operations but also that the orders are dual to each other. An easy example of an ordered additive group is the n-ple direct sum of $\mathbb{R}$ with the lexicographical order.

The primary importance of this generalization lies in its application to the theory of rings, but we are not dealing with topics of the sort.

LEMMA 5.7.1 Let v be an additive valuation of a field K. (i) If $0 \neq x \in K$, then $v(x^{-1}) = -vx$, (ii) $v1 = v(-1) = 0$, and (iii) $vx > vy$ implies $v(x + y) = vy$.

PROOF: Similar to Lemma 5.1.2. QED

Let v be an additive valuation of a field K. Just as in Lemma 5.2.2, $R_v = \{x \in K \mid vx \geq 0\}$ is a ring and $P_v = \{x \mid vx > 0\}$ is its unique maximal ideal. R_v is called the <u>valuation ring</u> of v, and P_v the <u>valuation ideal</u> of v. The field R_v/P_v is called the <u>residue class field</u> of v. Krull dim R_v is called the <u>rank</u> of v. Two additive valuations v and w are said to be <u>equivalent</u> to each other if their valuation rings are the same.

A subring R of a field K is called a <u>valuation ring</u> of K if there is a valuation v such that R is the valuation ring of v. By our definition of equivalence, such v is unique within equivalence.

THEOREM 5.7.2 For a subring R of a field K, the following six conditions (1) to (6) are equivalent to each other:

(1) R is a valuation ring of K.

(2) $x \in K$, $x \notin R$ imply $x^{-1} \in R$.

(3) K is the field of fractions of R, and it holds that if $a \notin bR$ $(a,b \in R)$, then $b \in aR$.

(4) K is the field of fractions of R; R has only one maximal ideal; and every finitely generated ideal of R is principal.

(5) R has only one maximal ideal, and every subring S of K properly containing R contains the inverse of some nonunit of R.

(6) Every ring S, such that $R \subseteq S \subseteq K$, coincides with $R_P = \{a/b \mid a,b \in R,\ b \notin P\}$ where P is a certain prime ideal of R.

PROOF: (1) $\Rightarrow$ (2): $x \in K$, $x \notin R$ imply $vx < 0$ (v being a valuation whose ring is R). Then $v(x^{-1}) > 0$ and $x^{-1} \in R$.

(2) $\Rightarrow$ (1): Assume that $x,y \in K$ and $xR \not\subseteq yR$. Then $x/y \notin R$ and $y/x \in R$. This implys that $yR \subseteq xR$. Now let G be $\{xR \mid 0 \neq x \in K\}$. Then G is a multiplicative group linearly ordered by the

containment relation. We write the multiplication by addition, and we have a new group $H = \{[xR] \mid xR \in G\}$ in which $[xR] + [yR] = [xyR]$. We introduce an order in H by $[xR] \geq [yR]$ if and only if $xR \subseteq yR$. Then H is an ordered additive group. The mapping v, such that $vx = [xR]$ $(x \neq 0)$, $v0 = \infty$, defines an additive valuation of K as is easily seen, and R is the valuation ring of v.

(2) ⟹ (3): The first half is obvious; the last part was proved above.

(3) ⟹ (2): If $x \in K$, $x \notin R$, then $x = a/b$ $(a,b \in R)$ and $a \notin bR$. Then $b \in aR$ and $b/a \in R$.

Thus we have proved the equivalence of (1) to (3).

(4) ⟹ (3): For $a,b \in R$, let c be such that $aR + bR = cR$. Set $x = a/c$, $y = b/c$. Then $xR + yR = R$. Since R has only one maximal ideal, it follows that one of x, y, say x, is a unit in R. Then $aR + bR = aR$ and $b \in aR$.

The converse [(1), (2), (3) ⟹ (4)] is obvious.

(1) ⟹ (6): If S is a ring in between R and K, then S satisfies condition (2) and S is a valuation ring of K; hence S has only one maximal ideal Q. Set $S' = R_{Q \cap R}$. Then $R \subseteq S' \subseteq S$ and therefore, by the same reasoning, S' is also a valuation ring of K. If $x \in S$, $x \notin S'$, then $x^{-1} \in S' \subseteq S$ and x^{-1} is a unit in S; x^{-1} is not a unit in S'. The maximal ideal of S' is $(Q \cap R)S' \subseteq Q \cap S'$, and therefore $x^{-1} \in Q$ which contradicts that x^{-1} is a unit in S. Thus (6) follows from (1).

(6) ⟹ (5): Very easy.

(5) ⟹ (3): Let R^* be the integral closure of R in K. Then Condition (5) and Corollary 4.7.5 show that $R^* = R$. Thus R is a normal ring. Similarly, we see that K is the field of fractions of R. Let $a,b \in R$ and $a \notin bR \neq 0$. By our assumption, there is a nonunit d of R such that $d^{-1} \in R[a/b]$, and we write $d^{-1} = c_0(a/b)^m + \cdots + c_m$ $(c_i \in R)$. Then $b^m = dc_0a^m + dc_1a^{m-1}b + \cdots + dc_mb^m$. Since R has only one maximal ideal, $1 - dc_m$ is a unit in R. Multiplying $(1 - dc_m)^{-1}$, we have $b^m = dc_0'a^m + dc_1'a^{m-1}b + \cdots + dc_{m-1}'ab^{m-1}$ $(c_i' \in R)$. This shows that b/a is integral over R. Since

R is normal, we have $b/a \in R$ and $b \in aR$. QED

COROLLARY 5.7.3 A valuation ring R of a field K is a normal ring. (i) If S is an intermediate ring between R and K, then S is a valuation ring and of the form R_P with a prime ideal P of R. (ii) The set of ideals in R is linearly ordered by the containment relation. (iii) If L is a subfield of K, then $R \cap L$ is a valuation ring of L.

PROOF: The first statement and (i) were proved already. If A, B are ideals of R and if $A \nsubseteq B$, then there is an element a of A which is not in B. Then, for any $b \in B$, we have $b \in aR$ and $B \subseteq aR \subseteq A$. (iii) follows from (2) in Theorem 5.7.2. QED

COROLLARY 5.7.4 If P is a prime ideal of a valuation ring R, then $P = PR_P$ and R/P is a valuation ring of R_P/P.

PROOF: If $a \in PR_P$ and $a \notin R$, then $a^{-1} \in R \subseteq R_P$ and a is a unit in R_P which contradicts $a \in PR_P$. Thus $PR_P \subseteq R$ and $P = PR_P \cap R = PR_P$. Assume next that $a^*, b^* \in R/P$. Let a, b be representatives of a^*, b^*. Then $aR \subseteq bR$ or $bR \subseteq aR$. It follows that $a^*(R/P) = aR/P \overset{\subseteq}{\underset{\supseteq}{}} bR/P = b^*(R/P)$. QED

We explicitly state here the relationship between nonarchimedean valuations and additive valuations (cf. our statement after the definition of additive valuations).

THEOREM 5.7.5 If v is a nontrivial nonarchimedean valuation of a field K, then, fixing a real number r such that $1 > r > 0$, we have an additive valuation w of rank 1 by defining $wx = \log_r vx$. Conversely, if w is an additive valuation of rank 1, then there is an injection f of the value group G of w into the additive group $\mathbf{R}$ of real numbers (f is not just an injection of a group; f preserves order), and we have a nontrivial nonarchimedean valuation v by defining $vx = r^{f(wx)}$. In each case, the valuation ring of v coincides with that of w. Therefore, the correspondence gives a one-one correspondence between the equivalence classes of nontrivial nonarchimedean valuations and the equivalence classes of additive valuations of rank 1.

PROOF: As for the first assertion, it suffices to show that w is of rank 1. Let a be such that $0 \neq va < 1$. a is a nonunit in the valuation ring R of v. If b ($\neq 0$) is another nonunit of R, then $0 \neq vb < 1$, and there are natural numbers s, t such that $(va)^s < vb$, $(vb)^t < va$. Then $a^s \in bR$, $b^t \in aR$. Therefore Krull dim R = 1. Conversely, assume that w is an additive valuation of rank 1 with value group G, valuation ring R, and valuation ideal P. Take an element a such that $0 \neq a \in P$. For each natural number n and an element b of K, let $e = e(n,b)$ be such that $b^n \notin a^e R$, $b^n \in a^{e-1}R$. (The existence follows from the fact that ht P = 1.) This means that $(e - 1)(wa) \leq n(wb) < e(wa)$. If $m \geq n$, then $|(e(n,b)/n) - (e(m,b))/m| < 1/n$ and we see the existence of $\lim e(n,b)/n$. Let $u(b) = \lim e(n,b)/n$; the value $u(b)$ is determined by wb, and therefore f such that $f(wb) = u(b)$ is a well-defined mapping of the value group G of w into $\mathbb{R}$. If $g,h \in G$, then $g = wb$, $h = wc$ with $b, c \in K$ and $g + h = w(bc)$. $b^n \notin a^{e(n,b)}R$, $b^n \in a^{e(n,b)-1}R$, $c^n \notin a^{e(n,c)}R$, $c^n \in a^{e(n,c)-1}R$ imply $(bc)^n \notin a^{e(n,b)+e(n,c)}R$, $(bc)^n \in a^{e(n,b)+e(n,c)-2}R$. Thus $e(n,b) + e(n,c) - e(n,bc) = 0$ or 1, and we have $f(g + h) = fg + fh$. Thus f is a group homomorphism. If $g > 0$, then $e(n,b) > 1$ for a large n, $(e(m,b) - 1)/m$ is monotone decreasing for $m = n, n^2, \ldots$ and we see that $\lim e(m,b)/m = \lim_t (e(n^t,b) - 1)/n^t \geq (e(n,b) - 1)/n > 0$. Thus $g > 0$ implies $fg > 0$. Thus f is injective, and furthermore it follows that if $g > h$, then $f(g - h) > 0$ and $fg > fh$. Thus f is an injection of an ordered group. Other assertions in the theorem are easily proved. QED

Thus we saw that an additive valuation of rank one is an expression of a nonarchimedean valuation. Therefore, if an additive valuation w corresponds to a p-adic valuation, then w is also called a p-*adic* (additive) valuation.

In constructing valuation rings, it is worthwhile to note the following:

THEOREM 5.7.6 Let P be the maximal ideal of a valuation ring R of a field K. If R* is a valuation ring of the field R/P, then

$S = \{x \in R \mid (x \text{ modulo } P) \in R^*\}$ forms a valuation ring of K. P is a prime ideal of S, $S/P \cong R^*$, and $S_P = R$. In particular, Krull dim S = Krull dim R + Krull dim R^*.

This S is called the composite of R with R^*.

PROOF: Assume that $a \in K$, $a \notin S$. (i) If $a \notin R$, then $a^{-1} \in P \subset S$. (ii) If $a \in R$, then $(a \text{ modulo } P) \notin R^*$ and $(a^{-1} \text{ modulo } P) \in R^*$. Thus, in any case, $a^{-1} \in S$, and S is a valuation ring of K. Since $S \subseteq R \subseteq K$, we see that $R = S_{P \cap S}$ [Theorem 5.7.2, (6)]. Since $P \cap S = P$, we have $R = S_P$. The other assertions in the theorem are easy. QED

We add here the notion of a place. A place f of a field K is a mapping of K into $L \cup \{\infty\}$ (where L is a field) such that $f1 = 1$ and such that $a,b \in K$ imply $f(ab) = (fa)(fb)$, $f(a + b) = fa + ab$; here the symbol ∞ is handled under the following convention: (i) If $0 \neq c \in L$, then $c\infty = \infty c = \infty\infty = \infty$. (ii) If $d \in L$, then $d + \infty = \infty + d = \infty$. (iii) $\infty + \infty$, 0∞, and $\infty 0$ are indefinite. (iv) $\infty^{-1} = 0$. Note that if $fx = \infty$, then $f(x^{-1})fx = f(x^{-1}x) = 1$, and $f(x^{-1}) = \infty^{-1} = 0$ by (iv). Two places f, f' of a field K are said to be equivalent to each other if the mapping $fa \rightsquigarrow f'a$ gives an isomorphism.

THEOREM 5.7.7 **There is** a one-one correspondence between the set of equivalence classes of places of a field K and the set of valuation rings of K such that: If a place f corresponds to a valution ring R, then $R = \{x \in K \mid fx \neq \infty\}$; the restriction of f on R is equivalent to the natural mapping Ψ of R onto R/P (P being the maximal ideal of R), namely, $x \rightsquigarrow fx$ gives an isomorphism of fR to R/P.

PROOF: It is obvious that a valuation ring R defines a place f. Conversely, let f be an arbitrary place of K. Set $R = \{x \mid fx \neq \infty\}$. R is a ring. If $x \in K$, $x \notin R$, then $fx = \infty$ and $f(x^{-1}) = \infty^{-1} = 0$ which implies that $x^{-1} \in R$. Thus R is a valuation ring. If we restrict f to R, then the kernel is the valuation ideal, and therefore f is equivalent to the place corresponding to R. QED

5.8 APPROXIMATION THEOREM

We begin with the approximation theorem of multiplicative valuations:

THEOREM 5.8.1 Let $v_1, \ldots, v_n$ be mutually nonequivalent, nontrivial multiplicative valuations of a field K. If $r_1, \ldots, r_n$ are positive real numbers and if $a_1, \ldots, a_n$ are elements of K, then there is an element b of K such that $v_i(b - a_i) < r_i$ for all i.

For the proof, we need:

LEMMA 5.8.2 If a multiplicative valuation v of a field K satisfies the triangular inequality, then for $c \in K$, we have:

(i) $v(c/(1 + c)) \le 1 + v(1 + c)^{-1}$.

(ii) $vc > 2$ implies $1/2 < v(c/(1 + c)) < 2$.

(iii) $vc > 1/2$ implies $v(1/(1 + c)) < 2$.

PROOF: (i): $c/(1 + c) = 1 - (1 + c)^{-1}$. (ii): $1 < v(1 + c) \le vc + 1 > 3$, and $1/3 > (vc + 1)^{-1} \le v(1 + c)^{-1} < 1$. Therefore $1/2 < 1 - (1/3) < vc/(vc + 1) \le v(c/(1 + c)) \le v(1 + c)^{-1} + 1$ [by (i)], and this is less than 2 by the inequality just obtained. (iii): $v(1 + c) \ge 1 - vc > 1/2$. QED

PROOF of THEOREM 5.8.1: We may assume that every v_i satisfies the triangular inequality because we may replace each v_i with an equivalent valuation. We claim that there are $d_1, \ldots, d_n \in K$ such that $v_i(d_i) > 1$, $v_i(d_j) < 1$ $(i \ne j)$. By symmetricity, it suffices to show the existence of d_1. The case where $n = 2$: Theorem 5.1.3 shows that there are $c_1, c_2 \in K$ such that $v_1(c_1) > 1$, $v_2(c_1) \le 1$, $v_2(c_2) > 1$, $v_1(c_2) \le 1$. Then $d_1 = c_1 c_2^{-1}$ is the required element. The case where $n \ge 3$: By induction on n, there are $c_1, c_2 \in K$ such that $v_1(c_1) > 1$, $v_2(c_1) < 1, \ldots, v_{n-1}(c_1) < 1$; $v_1(c_2) > 1$, $v_3(c_2) < 1, \ldots, v_n(c_2) < 1$. If either $v_n(c_1) < 1$ or $v_2(c_2) < 1$, then we can take d_1 to be c_1 or c_2. Therefore we assume that $v_n(c_1) \ge 1$ and $v_2(c_2) \ge 1$. If the equality holds in both, then we may take $d_1 = c_1 c_2$. Therefore, by the symmetricity between 2 and n, we may assume that $v_2(c_2) > 1$, $v_2(c_2) \ge v_n(c_1) \ge 1$. Considering powers of c_i if necessary, we may assume that $v_1(c_i) > 2$, $v_j(c_i)$

$\leq 1/2$ $(i = 1, 2;\ j = 3, \ldots, n-1)$, $v_2(c_1) < 1/2$, $v_2(c_2) > 2$, $v_n(c_2) \leq v_n(c_1c_2) < 1/2$. Set $d_1 = c_1c_2/(1 + c_2)$. Then Lemma 5.8.2 shows that this d_1 is the required element. Let t be a large natural number such that $v_i(d_i^t) > r_i^{-1} + 1$, $v_j(d_i^t) < r_i/2$ $(i \neq j)$, and set $e_i = d_i^t/(1 + d_i^t)$. Then we see that $v_i(e_i - 1) < r_i$, $v_j(e_i) < r_j$ $(i \neq j)$. Now, take positive real numbers s_i such that $\sum_j v_i(a_j)s_i < r_i$ and obtain e_i as above with respect to s_i instead of r_i. Then set $b = \sum a_ie_i$. Then $v_i(b - a_i) = v_i(a_i(e_i - 1) + \sum_{j\neq i} a_je_j) \leq v_i(a_i(e_i - 1)) + \sum_{j\neq i} v_i(a_je_j) < v_i(a_i)s_i + \sum_{j\neq i} v_i(a_j)s_i < r_i$. QED

At the next step, we consider additive valuations $v_1, \ldots, v_n$. If they are of rank 1, then the theorem above can be applied. But, if there are v_i of rank at least 2, then the immediate adaption of the theorem become false. For instance, consider composites R_1, R_2 of a valuation ring R with valuation rings R', R'' of the residue class field. In this case if a_1 is a nonunit of R, there is no element b such that both $b - a_1$ and $b - a_1^{-1}$ are every close to zero along R_1 and R_2, respectively.

In order to discuss the approximation theorem for additive valuations, we begin with a preliminary.

LEMMA 5.8.3 Let $R_1, \ldots, R_n$ be valuation rings of a field K such that $R_i \not\subseteq R_j$ if $i \neq j$. Set $D = \cap R_i$. Then for each $a \in K$, there is a natural number $s \geq 2$ such that both $x = (1 + a + \cdots + a^{s-1})^{-1}$ and $y = a/(1 + a + \cdots + a^{s-1})$ are in D.

PROOF: Let v_i be a valuation defined by R_i. (i) If $a \notin R_i$, then $v_ia < 0$ and $v_i(1 + \cdots + a^{s-1}) = v_i(a^{s-1})$. Then $v_ix = (1 - s)v_ia < 0$, $v_iy = (2 - s)v_ia < 0$. Thus $x,y \in R_i$. (ii) If $v_ia > 0$, then $1 + a + \cdots + a^{s-1}$ is a unit in R_i, and therefore $x,y \in R_i$. (iii) If $v_ia = 0$ and $v_i(a^t - 1) = 0$ for every natural number t, then since $(1 - a)(1 + a + \cdots + a^{s-1}) = 1 - a^s$, we see that $1 + a + \cdots + a^{s-1}$ is a unit in R_i and $x,y \in R_i$. (iv) Assume that $v_i(a - 1) > 0$. Then $x,y \in R_i$ if s is not a multiple of the characteristic of the residue class field of R_i. (v) Assume that $v_i(a - 1) = 0$ and $v_i(a^t - 1) > 0$ for a t. Let m be the minimum of such t. Then other t are multiples of m. Therefore $x, y \in R_i$ if s is not a multiple of m. Thus, if s is not a multiple of a certain natural number > 1, then $x, y \in R_i$ for all i and $x, y \in D$. QED

Next, we state the <u>independence theorem of additive valuations</u>:

THEOREM 5.8.4 Let $R_1, \ldots, R_n$ be valuation rings of a field K such that $R_i \not\subseteq R_j$ if $i \neq j$. Let P_i be the maximal ideal of R_i, and set $D = \cap R_i$, $Q_i = P_i \cap D$. Then (i) $R_i = D_{Q_i}$ and (ii) the maximal ideal of D are exactly $Q_1, \ldots, Q_n$.

PROOF: Since $D \subseteq R_i$, we have $D_{Q_i} \subseteq R_i$. Let $a \in R_i$. Then with $s \geq 2$, we have x, y in Lemma 5.8.3, which are in D. Since $x^{-1} = 1 + a + \cdots + a^{s-1} \in R_i$, we see that $x \notin Q_i$, and therefore $a = y/x \in D_{Q_i}$. Thus (i) is proved, and we see that $Q_i \not\subseteq Q_j$ because $R_j \not\subseteq R_i$. Thus, in order to prove (ii), it suffices to show that if M is a maximal ideal of D, then there is j such that $M \subseteq Q_j$. Assume that $M \not\subseteq Q_i$ for every i. Then there is $a_i \in M$ which is not in Q_i for each i. Furthermore, there is an element b_i of $\cap_{j \neq i} Q_j$ which is not in Q_i (Lemma 2.4.5). Set $c = \sum a_i b_i$. Then $c \notin Q_i$ for every i; hence c is a unit in every R_i. Thus $c^{-1} \in D$ and c is a unit in D, which contradicts $c \in M$. QED

COROLLARY 5.8.5 Under the same notation as above, if $A_1, \ldots, A_n$ are ideals of $R_1, \ldots, R_n$, respectively, and if $\sqrt{A_i} \not\subseteq R_j$ for any $i \neq j$, then $D/(\cap A_i) \cong R_1/A_1 \oplus \cdots \oplus R_n/A_n$.

PROOF: Set $B_i = A_i \cap D$. Assume that there is a maximal ideal Q_s containing $B_i + B_j$ $(i \neq j)$. Then $D_{\sqrt{B_i}} \supseteq D_{Q_i} = R_i$ and therefore $D_{\sqrt{B_i}} = (R_i)_{\sqrt{A_i}}$. Similarly, since $D_{\sqrt{B_i}} \supseteq D_{Q_s} = R_s$, we see that $\sqrt{A_i}$ is an ideal of R_s. Thus $i = s$. Similarly, $j = s$ and $i = j$, a contradiction. Therefore $D/(\cap A_i) = D/B_1 \oplus \cdots \oplus D/B_n$ (Theorem 2.3.1). D/B_i has a unique maximal ideal Q_i/B_i, and if $x \in D$, $x \notin Q_i$, then the class of x in D/B_i is a unit. Therefore $D/B_i \cong D_{Q_i}/B_i D_{Q_i} = R_i/A_i$. QED

From this, we obtain the following approximation theorem:

THEOREM 5.8.6 Let $R_1, \ldots, R_n$ be valuation rings of additive valuations $v_1, \ldots, v_n$ of a field K. Assume that $R_i \not\subseteq R_j$ if $i \neq j$. Let T_{ij} be the largest among common prime ideals of R_i and R_j. † Let T_i be the largest of T_{ij} $(j \neq i)$. If $a_1, \ldots, a_n, b_1, \ldots, b_n \in K$ are such that (i) $b_i \in R_i$, $b_i \notin T_i$ and (ii) there are natural numbers t_i such that $v_i(b_i^{t_i} a_i) > 0$, then there is an element $c \in K$ such that $v_i(c - a_i) > v_i(b_i)$ for every i.

PROOF: We may assume that each b_i is a nonunit in R_i. (1) If $a_i \in R_i$ for every i, then we apply Corollary 5.8.5 to $D/(\cap b_i^2 R_i) = R_1/b_1^2 R_1 \oplus \cdots \oplus R_n/b_n^2 R_n$. Let c be a representative in D of $\sum (a_i$ modulo $b_i^2 R_i)$. Then $v_i(c - a_i) \geq v_i(b_i^2) > v_i(b_i)$. (2) The general case: Apply (1) to the case where both a_i, b_i are $b_i^{t_i}$, and we have an element $d \in D$ such that $v_i(d - b_i^{t_i}) > v_i(b_i^{t_i})$. Then $d \neq 0$ and $v_i(d) = v_i(b_i^{t_i})$. Then apply (1) to $a_i d$ and $b_i d$, and we have $e \in D$ such that $v_i(e - a_i d) > v_i(b_i d)$. Then $c = e/d$ is the required element. QED

5.9 EXTENSIONS OF A VALUATION

In this section, we aim to prove results related to the following:

THEOREM 5.9.1 Let v be an additive or multiplicative valuation of a field K, and let L be an algebraic extension field of K. Let K^* be the algebraic closure of K. (1) v can be extended to a valuation w of K^*. (2) A valuation v' of L is an extension of v if and only if there is a K-injection f of L into K^* such that $v'x = w(fx)$. (3) If $[L : K]$ is finite, then the number of such extensions v' is at most $[L : K]_s$; if furthermore v is additive or nonachimedean, then the number is equal to the number of maximal ideals of the integral closure D in L of the valuation ring R_v of v.

† The existence of T_{ij} is seen by (i) ideals in a valuation ring are linearly ordered and (ii) the union of linearly ordered (under containment relation) prime ideals is a prime ideal.

Let us begin with the following theorem:

THEOREM 5.9.2 Let S be a subring of a field K, and let $0 \subset P_1 \subset P_2 \subset \cdots \subset P_n$ be a chain of prime ideals in S. Then there is a valuation ring R of K having prime ideals Q_i lying over P_i for $i = 1, \ldots, n$.

PROOF: If $n = 0$, then we may set $R = K$. Assume that $n = 1$. Considering S_P, we may assume that $P = P_1$ is the unique maximal ideal of S. Consider the set F of subrings T of K such that $S \subseteq T$ and $1 \notin PT$. It is easily seen that this is an inductive set, and therefore there is a maximal member R in F. R is integrally closed in K (otherwise, contradiction by Corollary 4.7.5) and R has only one maximal ideal M (otherwise, $R_M \in F$). Let $x \in K$, $x \notin R$. Then $PR[x]$ contains 1 and $1 = c_0 + c_1x + \cdots + c_sx^s$ ($c_i \in PR$). Since M is the unique maximal ideal and contains P, we see that $1 - c_0$ is a unit in R, and multiplying $(1 - c_0)^{-1}$, we may assume that $c_0 = 0$. Then the relation shows that x^{-1} is integral over R. Since R is integrally closed in K, we have $x^{-1} \in R$. This proves that R is a valuation ring. $M \supseteq P$ and P is maximal; hence M lies over P. Assume now that $n > 1$. By induction on n, we see the existence of a valuation ring V of K having prime ideals $Q_1 \subset \cdots \subset Q_{n-1}$ such that each Q_i lies over P_i. We may assume that Q_{n-1} is maximal. S/P_{n-1} is a subring of the field V/Q_{n-1}. Therefore, by the case $n = 1$, there is a valuation ring V^* of V/Q_{n-1} which has a prime ideal lying over P_n/P_{n-1}. Then the composite R of V with V^* is the required ring. QED

Let R be a valuation ring of a field K, and let L be an extension field of K. A valuation ring R' of L is said to be an <u>extension</u> of R if $R' \cap K = R$. Note that this is equivalent to the fact that R' is the valuation ring of a suitable extension of a valuation given by R.

THEOREM 5.9.3 Let R, K, L be as above, and assume that L is algebraic over K. Let R^* be the integral closure of R in L. Then R' is a valuation ring of L which is an extension of R if and only if $R' = R^*_{M^*}$ with a maximal ideal M^* of R^*.

PROOF: (1) Assume first that L is normal and of finite degree over K. Theorem 5.9.2 shows that there is an extension R' (cf. Exercise 5.7.4). Let G be the Galois group $G(L/K)$, set $D = \cap_{g\in G} R'^g$, and let $R_1, \ldots, R_n$ be all of mutually distinct R'^g. Then $R_i \not\subseteq R_j$ if $i \neq j$, because R_i and R_j are conjugate to each other. Let $d \in D$, and let $f(X) = X^t + c_1X^{t-1} + \cdots + c_t$ $(c_i \in K)$ be the minimal polynomial for d over K. Each c_j is a polynomial in the conjugates d^g of d. On the other hand, $d^g \in D$ implies that $v_i(d^g) \geq 0$ for a valuation v_i defined by R_i. Therefore $v_i(c_j) \geq 0$ and $c_j \in R_i \cap K = R$. Thus d is integral over R. Since D is normal, it follows that $D = R^*$. Theorem 5.8.4 shows that each R_i is of the form $R^*_{M^*}$, and conversely. Assume next that R'' is another extension of R to L. Then $D = \cap R''^g$ and $R'' = R_i$ for some i. This proves the present case.

(2) Assume that $[L : K] < \infty$. Let L^* be a finite normal extension of K containing L, and let T be the integral closure of R in L^*. Then V is an extension of R in L^* if and only if $V = T_M$ with a maximal ideal M of T by (1) above. If R' is an extension of R to L, then there is an extension V^* of R' to L^*. Then V^* is one of T_M. Thus $R' = T_M \cap L \supseteq R^*_{M\cap R^*}$. Since $R^* = T \cap L = \cap_M (T_M \cap L)$ and each $T_M \cap L$ is a valuation ring, it follows that $R^*_{M\cap R^*}$ is a valuation ring by Theorem 5.8.4, and therefore $R' = R^*_{M\cap R^*}$. Since T/M is algebraic over $R/(M \cap R)$, it follows that $M \cap R^*$ is a maximal ideal of R^*. It is obvious that each $R^*_{M^*}$ gives an extension of R, and we complete this case.

(3) The general case: Consider first $R^*_{M^*}$. For $x \in L$, we set $R'' = R^* \cap K(x)$ and $V = R''_{M^*\cap R''}$. Then V is a valuation ring, and either x or x^{-1} is in V. Since $R^*_{M^*}$ contains V, it follows that $R^*_{M^*}$ is a valuation ring of L. $M^* \cap R$ is a maximal ideal of R, and therefore we see that $R^*_{M^*}$ is an extension of R. Conversely, if R' is an extension of R in L, then letting P' be the maximal ideal of R', we set $M' = P' \cap R^*$. Then for each $a \in R'$, $a \in K(a) \cap R' = (R^* \cap K(a))_{M'\cap K(a)} \subseteq R^*_{M'}$. Thus $R' \subseteq R^*_{M'}$.

Since the converse inclusion is obvious, we have $R' = R^*_{M'}$. QED

COROLLARY 5.9.4 If an additive valuation v of a field K is extended to a valuation w of an algebraic extension L, then the rank of w is equal to the rank of v.

PROOF: This follows from Theorem 4.8.2. QED

COROLLARY 5.9.5 Let v be an additive or nonarchimedean valuation of a field K, and let L be an arbitrary extension field of K. Then v is extended to a valuation w of L. In the case of an additive valuation, we can extend it so that the rank of w coincides with that of v.

PROOF: The nonarchimedean case is regarded as an additive case of rank ≤ 1, and we prove the additive case only. Let $T = \{t_\lambda\}$ be a transcendence base of L over K, and let F be the ring generated by these t_λ over the valuation ring R of v. Let P be the maximal ideal of R. Since F is regarded as a polynomial ring over R, PF is a prime ideal and we can consider $V = F_{PF}$. Each element of V is expressed in the form af/g with $a \in R$, $f,g \in F$ such that a coefficient (as a polynomial in t_λ) of each of f and g is 1. Then, such f/g is a unit in V. Therefore, we see easily that V is a valuation ring (Krull dim V = Krull dim R) and is an extension of R. Now, V is extended to a valuation ring of L, and the assertion follows from Corollary 5.9.4. QED

PROOF of THEOREM 5.9.1: We first assume that v is an additive valuation. The existence of w follows from Theorem 5.9.3. As for (2), it suffices to prove the case where $L = K^*$. Let R^* be the integral closure of the valuation ring R of v in K^*. Then extensions of v are given by $R^*_{M^*}$ with maximal ideals M^* of R^*. Since the maximal ideals of R^* are conjugates to each other (Theorem 4.7.7), we see the assertion easily. (3) is a special case of (2). The nonarchimedean case can be regarded as an additive case of rank ≤ 1. Now we assume that v is an archimedean valuation. By Theorem 5.6.1, we may assume that K is a subfield of the complex number field C and that $vx = |x|$ for every $x \in K$. Then $K^* \subseteq C$ and

w such that $wx = |x|$ gives an extension of v to K^*. If w' is an extension of v in K^*, then Theorem 5.6.1 shows that $w'x = |fx|^t$ with an injection f of K^* in $\mathbb{C}$ and a positive number t. Since $w'x = |x|$ for $x \in K$, t must be 1, and we complete the proof. QED

THEOREM 5.9.6 Let v be an additive valuation of a field K, and let L be an extension field of K of finite degree over K. Let $w_1, \ldots, w_n$ be all of extensions of v to L. Let the valuation rings, the valuation ideals, and the value groups of v, w_i be R, T_i; M, P_i and G, H_i, respectively. Set $e_i = \#(G/H_i)$, $f_i = [T_i/P_i : R/M]$. Then they are finite, and it holds that

$$\textstyle\sum_i e_i f_i \leq [L : K].$$

These e_i and f_i are called the ramification exponent and the degree of w_i over v. If $e_i = 1$, we say that w_i is unramified over v.

PROOF: (1) We first prove that $e_i f_i \leq [L : K]$ for each i. Let $a_j \in T_i$ be such that $c_j = (a_j \text{ modulo } P_i)$ $(j = 1, \ldots, s)$ are linearly independent over R/M, and let $\gamma_k \in H_i$ $(k = 1, \ldots, t)$ be such that their residue classes modulo G are distinct from each other. Let $b_k \in L$ be such that $w_i(b_k) = \gamma_k$. Assume that $\sum d_{jk}a_j b_k = 0$ with $d_{jk} \in K$ (some $d_{jk} \neq 0$). Let I be the set of (j,k) such that $w_i(d_{jk}a_j b_k)$ is minimum among all of $w_i(d_{j'k'}a_{j'}b_{i'})$ $(j' \leq s,\ k' \leq t)$. Since $w_i(x) > w_i(y)$ implies $w_i(x + y) = w_i(y)$, we have $w_i(\sum_{(j,k)\in I} d_{jk}a_j b_k) > w_i(d_{jk}a_j b_k)$ $[(j,k) \in I]$. Thus $\#(I) \geq 2$. $w_i(d_{jk}a_j b_k) = w_i(d_{jk}) + \gamma_k \in G + \gamma_k$. Therefore k must be common to all $(j,k) \in I$. Then, dividing by one d_{jk}, we may assume that some $d_{jk} = 1$. Then other d_{jk} are units in R. Then $w_i(\sum_{(j,k)\in I} d_{jk}a_k) > 0$, which contradicts the linear independence of c_k. Therefore $a_j b_k$ $(j \leq s,\ k \leq t)$ are linearly independent over K, and we have $e_i f_i \leq [L : K]$. Note that this result implies in particular that e_i, f_i are all finite.

(2) If v is of rank zero, then $R = K$, $n = 1$, $e_1 = 1$, $f_1 = [L : K]$, and the assertion is obvious in this case.

(3) Assume that v is of rank 1. We may assume that $H_i \subseteq \mathbb{R}$ (the real number field). Let $a^*_{i1}, \ldots, a^*_{if_i}$ be a linearly independ-

ent base of T_i/P_i over R/M, and let $\beta_{i1} = 0, \beta_{i2}, \ldots, \beta_{ie_i}$ be a set of representatives for H_i/G. Let β be an element of G such that $-\beta < \beta_{ij} < \beta$ for all (i,j). By the approximation theorem, there are elements $a_{ij}, c_{ik} \in L$ such that (i) $a^*_{ij} = (a_{ij}$ modulo $P_i)$; $w_k(a_{ij}) > \beta$ if $k \neq i$, (ii) $w_i(c_{ij}) = \beta_{ij}$; $w_k(c_{ij}) > \beta$ if $k \neq i$. Assume that $\sum d_{ijk}a_{ij}c_{ik} = 0$ $(d_{ijk} \in K;$ some $d_{ijk} \neq 0)$. Set $m_i = \min_{tjk}\{w_i(d_{tjk}a_{tj}c_{tk})\}$, $m = \min_i\{m_i\}$. We may assume that $m_1 = m$ and $w_1(d_{tuv}a_{tu}c_{tv}) = m$. Then $t = 1$; for, otherwise, $v(d_{tuv}) + 2\beta < w_1(d_{tuv}a_{tu}c_{tv}) \leq w_t(d_{tuv}a_{tu}c_{tv}) = v(d_{tuv}) + \beta_{tv} < v(d_{tuv}) + \beta$, which contradicts $\beta > 0$. Set $I = \{(i,j,k) \mid w_i(d_{ijk}d_{ij}c_{ik}) = m\}$. Then we have seen that $(i,j,k) \in I$ implies that $i = 1$. On the other hand, since $\sum d_{ijk}a_{ij}c_{ik} = 0$, we have, as in (1) above, $w_1(\sum_{(ijk)\in I} d_{ijk}a_{ij}c_{ik}) > m$, and we have a contradiction just as in (1). This completes the proof of the rank 1 case.

(4) We consider the case where v is of a finite rank $r > 1$. Let Q be the prime ideal of height 1 in R. Note that the integral closure D of R in L coincides with $\cap T_i$ (Theorem 5.9.3). Set $S = R - Q$. Then D_S is the integral closure of R_Q in L, and therefore D_S is the intersection of valuation rings U_i which give extensions of R_Q in L. On the other hand, $D_S = \cap(T_i)_S$, and therefore $U_i = (T_j)_S$ for some (i,j) (for each i or for each j). Thus, letting Q_i be the prime ideal of height 1 in T_i, we may assume that (i) $U_1, \ldots, U_s$ are all of mutually distinct U_i, and (ii) $U_i = (T_i)_S = (T_i)_{Q_i}$ for $i \leq s$ [hence, for $j > s$, Q_j coincides with some Q_i $(i \leq s)$; cf. Corollary 5.7.4]. Let e'_i, f'_i be the ramification exponent and degree, respectively, of U_i over R_Q. Then, by the rank 1 case, we have $\sum e'_i f'_i \leq [L : K]$. For each U_j, let I_j be the set of i such that $(T_i)_S = U_j$. To such a pair (U_j, T_i), let e''_i, f''_i be the ramification exponent and degree, respectively, of T_i/Q_i over R/Q. Then $e_i = e'_j \cdot e''_i$, $f_i = f''_i$. By induction on r, we have $\sum_{i\in I_j} e''_i \cdot f''_i \leq f'_j$ and we have $\sum e_i f_i \leq [L : K]$.

(5) Lastly, we observe the general case. Let $L = K(y_1, \ldots, y_m)$. Let K_0 be a subfield of K such that all the y_i are algebraic

over K_0, and set $L_0 = K_0(y_1,\ldots,y_m)$. Let v' and w_i' be the restrictions of v, w on K_0, L_0, respectively. Among possible K_0, we choose one so that (i) if $i \neq j$, then $w_i' \neq w_j'$ (for the purpose, it suffices that L_0 contains an h_{ij} which is in P_i but not in P_j); (ii) some representative of each class of H_i/G is in the value group of w_i'; (iii) for each i, a set of representatives of a linearly independent base of T_i/P_i over R/M is contained in L_0; and (iv) $[L_0 : K_0] = [L : K]$. These conditions are satisfied by the fact that L_0 contains a finite number of suitable elements. Therefore we can choose such K_0 to be finitely generated over the prime field. Let e_i^*, f_i^* be the ramification exponent and degree of w_i' over v'. Then our construction shows that $e_i^* \geq e_i$, $f_i^* \geq f_i$. Since K_0 is finitely generated, (4) is applied (cf. Exercise 5.7.9) and we have $\sum e_i^* f_i^* \leq [L_0 : K_0] = [L : K]$, thus completing the proof. QED

THEOREM 5.9.7 Let K, L, e_i, f_i etc. be as above. Then:

(1) If v is <u>discrete</u> (i.e., the value group is cyclic) and if $D = \cap T_i$ is a finite R-module, then $\sum e_i f_i = [L : K]$.

(2) If v is discrete and if L is separably algebraic over K, then $\sum e_i f_i = [L : K]$.

(3) If the maximal ideal M of R is not principal and if D is a finite R-module, then $e_i = 1$ for all i and $\sum f_i = [L :K]$.

(4) If there is an element b of D such that (i) $L = K(b)$ and (ii) the discriminant d, of the minimal polynomial $g(X) = X^s + a_1X^{s-1} + \cdots + a_s$ for b over K, is not in the maximal ideal M of R, then $e_i = 1$ for all i and $\sum f_i = [L : K]$.

Before proving this, we prove the following result which is called the <u>lemma of Krull-Azumaya</u> or <u>of Nakayama</u>.

THEOREM 5.9.8 Let J be the <u>Jacobson radical</u> (i.e., the intersection of all maximal ideals) of a ring R. If M is a finite R-module, N a submodule and if $M = N + JM$, then it holds that $M = N$.

PROOF: (1) The case where $N = 0$: Let $m_1, \ldots, m_r$ be a set of generators for M. Since $M = JM$, $m_i = \sum c_{ij}m_j$ with $c_{ij} \in J$.

Lemma 2.9.1 shows that $dm_j = 0$ for all j where $d = \det(\delta_{ij} - c_{ij})$ (δ_{ij} is the Kronecker δ). Since $c_{ij} \in J$, we have $d - 1 \in J$ and d is a unit in R. Thus $m_j = 0$ for all j and $M = 0$.

(2) The general case: Consider $M^* = M/N$. Then M^* is a finite R-module and $M^* = JM^*$. Hence $M^* = 0$ by (1). Thus $M = N$. QED

PROOF of THEOREM 5.9.7: (1): If $R = K$, then the assertion is obvious and we assume that $R \neq K$. Let γ (>0) be a generator of the value group of v. If $\beta \in H_i$, then $e_i\beta \in G$. This shows that there is a minimal element among positive elements in H_i and H_i is also cyclic generated by δ_i (>0) and $e_i\delta_i = \gamma$. Let $b_i \in L$ be such that $w_i(b_i) = \delta_i$, $w_j(b_i) > \gamma$ for $j \neq i$. Let $a_{i1}, \ldots, a_{if_i}$ be elements of D such that (i) these modulo P_i form a linearly independent base of T_i/P_i over R/M and (ii) $w_j(a_{ik}) > 0$ if $j \neq i$ (approximation theorem). By our proof of Theorem 5.9.6, we see that $b_i^t a_{ij}$ ($i = 1, \ldots, n$; $t = 0, 1, \ldots, e_i-1$; $j = 1, \ldots, f_i$) are linearly independent over K. Let N be the R-submodule generated by these elements. Then we see that $D = N + MD$. Thus $D = N$ by Theorem 5.9.8, and $\sum e_i f_i = [L : K]$.

(2): R is noetherian (Exercise 5.7.7) and normal (Corollary 5.7.3), and therefore D is finite over R (Corollary 4.10.3). Hence (2) follows from (1).

(3): Let $a_{i1}, \ldots, a_{if_i}$ be elements of D such that these modulo P_i form a linearly independent base of R/M and such that $w_j(a_{ik}) > 0$ if $j \neq i$. Then a_{ij} ($i = 1, \ldots, n$; $j = 1, \ldots, f_i$) are linearly independent over R. Let N be the R-module generated by these elements. Then $N + (\cap P_i) = D$. Since M is not principal, we have $MT_i = P_i$ and $\cap P_i = \cap MT_i = MD$. Since D is finite over R by our assumption, we have $N = D$ and $\sum f_i = [L : K]$.

(4): $dD \subseteq R[b] \subseteq D$ by Theorem 4.10.2. Since d is a unit in R, we have $D = R[b]$. Let $g^*(X)$ be $g(X)$ modulo M. Since $d \notin M$, $g^*(X)$ has no multiple root. Therefore if $g^*(X) = g_1^* \cdots g_t^*$ is a prime factorization, then $D/MD = R[X]/(MR[X] + g(X)R[X]) =$

$(R/M)[X]/g^*(X)(R/M)[X]$, which is the direct sum of $(R/M)[X]/g_i^*(R/M)[X]$ (Chinese remainder theorem). Therefore we have $e_i = 1$, $\sum f_i = [L : K]$. QED

5.10 PRODUCT FORMULA

We say that the <u>product formula</u> holds in a field K with respect to V if (i) $V = \{v_\lambda \mid \lambda \in \Lambda\}$ is a nonempty set of nontrivial multiplicative valuations of K such that if $\mu \neq \lambda$ then v_μ is not equivalent to v_λ and such that if $0 \neq x \in K$ then $\#\{\lambda \mid v_\lambda(x) \neq 1\}$ is finite and (ii) for every nonzero x in K it holds that $\Pi_{\lambda\in\Lambda} v_\lambda(x) = 1$ [note that the left-hand side is formally an infinite product but is actually a finite product because of our assumption (i)].

THEOREM 5.10.1 (1) In the rational number field Q, the product formula holds with respect to the following V: For each prime number p, let v_p be the p-adic valuation such that $v_p(p) = p^{-1}$. Let v_∞ be $|\ |$ (absolute value). Then $V = \{v_p \mid p$ is ∞ or a prime number$\}$.

(2) In the rational function field $K(x)$ in one variable x over a field K, the product formula holds with respect to the following V. Let r be a real number such that $0 < r < 1$. For each monic prime element p in $K[x]$, let v_p be the p-adic valuation such that $v_p(p) = r^{\deg p}$. Let v_∞ be the x^{-1}-adic valuation with respect to $K[x^{-1}]$ such that $v_\infty(x^{-1}) = r$. Then $V = \{v_p \mid p$ is ∞ or a monic prime element in $K[x]\}$.

PROOF: (1): If p is a prime number, then $v_p(x) \neq 1$ if and only if p divides either numerator or denominator of x and the validity of (i) follows from this. As for (ii), if q is a prime number, then $v_p(q) = 1$ if $\infty \neq p \neq q$; $v_q(q) = q^{-1}$, $v_\infty(q) = q$, and the product of them is 1. If $q = \pm 1$, then all $v_p(q) = 1$. Now, (ii) follows from these and $(\Pi\, v_p(a))(\Pi\, v_p(b)) = \Pi\, v_p(ab)$.

(2): Proof of this case is similar to (1) above, except for the fact that if $q = x^n + c_1x^{n-1} + \cdots + c_n$, then $v_\infty(x^n) > v_\infty(x^{n-1}) \geq v_\infty(c_1x^{n-1} + \cdots + c_n)$ and therefore $v_\infty(q) = v_\infty(x^n) = r^{-\deg q}$. QED

THEOREM 5.10.2 Assume that the product formula holds in a field K with respect to a set V of valuations. If L is a finite algebraic extension of K, then there is a set W of valuations of L such that (1) if $v \in V$ and if v' is an extension of v to L, then there is one (and only one) w in W which is equivalent to v; (2) if $w \in W$, then $w|_K$ is equivalent to some $v \in V$; and (3) the product formula holds in L with respect to W.

PROOF: If the assertion is true for an extension L' of L, then the restriction to L gives us the required result. Thus we may replace L with a suitable extension field. Thus we may assume that L is a normal extension of K. Let G be the Galois group. For each $v \in V$, let $v_1, \ldots, v_s$ be all the extensions of v to L; they are conjugates to each other and hence the number $f = \#\{g \in G \mid v_i^g = v_i\}$ is independent of i and $fs = \#(G)$. Let W_v be the set of v_i^f, and let W be the union of all W_v. Set $q = [L : K]_i$ $(fsq = [L : K])$. If $0 \neq x \in L$, then $\Pi_i v_i^{fq}(x) = (v_1(\Pi_{g\in G} x^g))^q = v((\Pi x^g)^q)$ [note that $(\Pi x^g)^q \in K$], and therefore $(\Pi_{w\in W} wx)^q = \Pi_{v\in V} v(\Pi x^g)^q = 1$. Therefore $\Pi_{w\in W} wx = 1$. QED

5.11 HENSEL LEMMA

We begin with the notion of the <u>resultant</u> of two polynomials $g(X) = c_0X^m + c_1X^{m-1} + \cdots + c_m$, $h(X) = d_0X^n + d_1X^{n-1} + \cdots + d_n$ over a ring R. It is defined to be the determinant of the following matrix $A(g,h)$;

$$A(g,h) = \begin{pmatrix} c_0 & c_1 & \ldots & c_m & 0 & \ldots & 0 \\ 0 & c_0 & c_1 & \ldots & c_{m-1} & c_m & 0 \ldots 0 \\ \ldots & \ldots & \ldots & \ldots & \ldots & \ldots & \ldots \\ 0 & \ldots 0 & c_0 & c_1 & \ldots & \ldots & c_m \\ d_0 & d_1 & \ldots d_{n-1} & d_n & 0 & \ldots & 0 \\ 0 & d_0 & \ldots & d_{n-1} & d_n & 0 \ldots & 0 \\ \ldots & \ldots & \ldots & \ldots & \ldots & \ldots & \ldots \\ 0 & \ldots & 0 & d_0 & d_1 & \ldots & d_n \end{pmatrix}$$

THEOREM 5.11.1 Under the notation as above, assume that R is an integral domain and that $c_0 d_0 \neq 0$.

(1) Let a_i be the $(i, n+m)$-cofactor of $A(g,h)$. (i) Assume that g is monic; then an element b of R is in $gR[X] + hR[X]$ if and only if there are elements $e_1, \ldots, e_{n+m}$, s, t of R such that $t \neq 0$, $sa_i = te_i$ for every i and such that $b = c_m e_n + d_n e_{n+m}$. (ii) The if part in (i) holds without assuming that g is monic.

(2) Let $\gamma_1, \ldots, \gamma_m$ and $\delta_1, \ldots, \delta_n$ be roots of $h(X)$ and $g(X)$ (in an algebraically closed field containing R), respectively. Then the resultant $\det(A(g,h))$ is equal to $\pm c_0{}^n d_0{}^m \cdot \Pi_{ij} (\gamma_i - \delta_j)$.

PROOF: (2): Let z, y, $u_1, \ldots, u_m$, $v_1, \ldots, v_n$ be algebraically independent elements over R, and consider $g^*(X) = \Pi (X - u_i)$, $h^*(X)) = \Pi (X - v_j)$ and zg^*, yh^* over $R^* = R[z,y,u_1,\ldots,u_m,v_1,\ldots,v_n]$. Then $\det(A(zg^*,yh^*)) = z^n y^m \cdot \det(A(g^*,h^*))$. Since the coefficients of g^*, h^* are $\pm$(elementary symmetric forms) in u_i, v_j, respectively, we see that the (i,j)-entry of $A(g^*,h^*)$ is homogeneous in $u_1, \ldots, u_m$, $v_1, \ldots, v_n$ and of degree $j - i$ (if $i \leq n$) or $j - (i - n)$ (if $i > n$). Therefore $\det(A(g^*,h^*))$ is homogeneous of degree mn. Take any i, j, and set $v_j = u_i$. Then g^* and h^* have a common root v_j. Then the system of linear homogeneous equations $A(g^*,h^*) \cdot {}^t(Y_{n+m},\ldots,Y_0) = 0$ has a nontrivial solution $(v_j{}^{n+m},\ldots,v_j,1)$, and therefore $A(g^*,h^*)$ must vanish in this case. Therefore by the factor theorem, $\det(A(g^*,h^*))$ (in the original case) is divisible by $v_j - u_i$. Let K be the field of fraction of R. Then polynomial rings over K are UFD, and therefore $\det(A(g^*,h^*))$ is divisible by $\Pi (v_j - u_i)$ as a polynomial over K. Compairing the degrees, we see that the ratio must be an element of K. Comparing the coefficients of $(\Pi v_j)^m$, we see that the ratio must be ± 1. Thus $\det(A(g^*,h^*)) = \pm \Pi (v_j - u_i)$ and $\det(A(zg^*,yh^*)) = \pm z^n y^m \cdot \Pi (v_j - u_i)$. This proves (2).

(1): Assume first that all a_i are 0. In this case, the resultant is zero, and therefore $\gamma_i = \delta_j$ for some (i,j). Then g, h have a nontrivial common factor $k(X)$ in $K[X]$. Then $(gR[X] +$

$hR[X]) \cap R \subseteq k(X)K[X] \cap R = 0$, and b must be 0. Therefore this case is proved ($e_i = 0$.)

Next we assume that some a_i are not zero. (i): Our condition for b is that $b = gk + hk'$ ($b \in R$; $k,k' \in R[X]$). Since g is monic, and since k' may be changed modulo g, we may assume that $\deg k' < \deg g$. Then $\deg gk = \deg hk' < \deg g + \deg h$, and therefore $\deg k < \deg h$. Write $k = e_1X^{n-1} + \cdots + e_{n-1}X + e_n$, $k' = e_{n+1}X^{m-1} + \cdots + e_{m+n}$. Coefficients of $X^{n-i}g$ form the i-th row of $A(g,h)$, and those of $X^{m-i}h$ form the $(n + i)$-th row. Therefore $d = gk + hk'$ is equivalent to the fact that $\sum e_i$(the i-th row of $A(g,h)) = (0,\ldots,0,b)$, namely that $(e_1,\ldots,e_{n+m})A(g,h) = (0,\ldots,0,b)$. If we regard this except for the last column as a system of linear equations on e_i, then $(a_1,\ldots,a_{n+m})$ is a solution [because $(a_1,\ldots,a_{n+m})A(g,h) = (0,\ldots,0,\det A(g,h))$] and other solutions are linearly dependent on this. Thus we have proved (i). (ii) is seen from our proof of (i) above. QED

COROLLARY 5.11.2 Under the notation as above, the resultant of g, h belongs to $(gR[X] + hR[X]) \cap R$.

LEMMA 5.11.3 Let d be the resultant of polynomials $g(X)$, $h(X)$ in X over a ring R. Assume that $g', h' \in R[X]$, $\deg g = \deg g'$, $\deg h = \deg h'$, $m \in R$ and that both $g - g'$ and $h - h'$ are in $dmR[X]$. Then the resultant d' of g', h' is divisible by d and $d - d' \in dmR$. In particular, if m is in the Jacobson radical of R, then $dR = d'R$.

PROOF: By the assumption and by the definition of resultants, we see that $d - d' \in dmR$, which implies that d' is divisible by d. So, $d'R \subseteq dR$ and $d'R + dmR = dR$, and the lemma of Krull-Azumaya proves the last statement. QED

We state here a Hensel lemma:

THEOREM 5.11.4 Let R be a complete valuation ring (namely, R is the valuation ring of a complete nonarchimedean valuation v) with maximal ideal P. Let $f(X)$, $g_0(X)$, $h_0(X)$ be polynomials in X over R such that (1) $g_0(X)$ is monic and (2) $f(X) - g_0(X)h_0(X)$

$\in d^2PR[X]$, where d is the resultant of $g_0(X)$, $h_0(X)$. Then there are $g(X), h(X) \in R[X]$ such that $g(X)$ is monic, both $g(X) - g_0(X)$ and $h(X) - h_0(X)$ are in $dPR[X]$, and $f(X) = g(X)h(X)$.

A special case where $d \notin P$ can be stated as follows:

COROLLARY 5.11.5 Let R, P, X be as above. Let f, g_0, $h_0 \in R[X]$ be such that (1) g_0 and h_0 are monic, (2) $f \equiv g_0h_0$ modulo P, and (3) (g_0 modulo P) and (h_0 modulo P) have no proper common factor. Then $f = gh$ with $g,h \in R[X]$ such that g is monic, $g - g_0 \in PR[X]$, and $h - h_0 \in PR[X]$. In particular, if $f = c_0X^n + c_1X^{n-1} + \cdots + c_n$ with $n \geq 2$, $c_i \in R$, $0 \neq c_0 \in P$, and $c_j \notin P$ for some j, then f is reducible.

PROOF: The first half follows from Theorem 5.11.4. The last follows from the first by the case where h_0 is a unit in R. QED

A more general form of the Hensel lemma will be given later as Theorem 5.11.14. As for Theorem 5.11.4, it follows from the following form of the Hensel lemma:

THEOREM 5.11.6 Let R, P, X be as above. Let $f, g_0, h_0 \in R[X]$ and $d \in R$ be such that (i) g_0 is monic, (ii) $d \in g_0R[X] + h_0R[X] + dPR[X]$, and (iii) $f - g_0h_0 \in d^2PR[X]$. Then there are $g,h \in R[X]$ such that (1) g is monic, (2) both $g - g_0$ and $h - h_0$ are in $dPR[X]$, and (3) $f = gh$.

PROOF: Let v be a nonarchimedean valuation whose ring is R. For a polynomial, say $k = c_0X^r + \cdots + c_r$, we understand by vk the maximum of $\{v(c_0),\ldots,v(c_r)\}$. Starting with (g_0,h_0), we want to construct a sequence of pairs (g_i,h_i) of polynomials and also a sequence of elements d_i of R so that $d_0 = d$, $v(d_i)$ decreases monotonously and tends to zero, every g_i is monic, and, for $i \geq 1$, $v(g_i - g_{i-1}) < v(d_{i-1})$, $v(h_i - h_{i-1}) < v(d_{i-1})$, $v(f - g_ih_i) \leq v(d_i)^2$, and $\deg h_i \leq \deg f - \deg g_0$. If we had such sequences, then writing $g_i = X^r + c_{i1}X^{r-1} + \cdots + c_{ir}$, $h_i = e_{i0}X^s + e_{i1}X^{s-1} + \cdots + e_{is}$ $(r = \deg g_0$, $s = \deg f - r)$, we see that $\{c_{ij} \mid i = 0, 1, \ldots\}$ and $\{e_{ik} \mid i = 0, 1, \ldots\}$ are Cauchy sequences, and hence they have limiting points c_j^* and e_k^*, respectively. Then with $g =$

$X^r + c_1^* X^{r-1} + \cdots + c_r^*$ and $h = e_0^* X^s + \cdots + e_s^*$, we have $v(f - gh) \le v(d_i^2)$ for all i and $f = gh$. Thus it suffices to prove the existence of these sequences. Take $e \in P$ such that $f - g_0 h_0 \in d^2 eR[X]$, and we set $d_i = de^i$. Assume that $g_0, h_0, \ldots, g_i, h_i$ are constructed, $f - g_i h_i \in d_i^2 R[X]$, and $f - g_i h_i = d^2 e^{2i} b_i k_i$ ($b_0 = e$, $b_i = 1$ for $i \ge 1$; $k_i \in R[X]$). By (ii), $d = g_i h_i' + h_i g_i' + de_i k_i'$ ($h_i', g_i', k_i' \in R[X]$; $e_i \in P$). If d in the right-hand side is replaced repeatedly by this form, then e_i is replaced by its power. Therefore we may assume that $e_i = e^2$. Now, $f - g_i h_i = de^{2i} b_i (g_i h_i' + h_i g_i' + de^2 k_i') k_i$. Since g_i is monic and of degree r, we can write $f - g_i h_i = de^{2i} b_i (g_i h_i'' + h_i g_i'' + de^2 k_i'')$ so that $\deg g_i'' < r$, $\deg k_i'' < r$. Set $g_{i+1} = g_i + de^{2i} b_i g_i''$, $h_{i+1} = h_i + de^{2i} b_i h_i''$. Then $f - g_{i+1} h_{i+1} = d^2 e^{2i+2} b_i k_i'' - d^2 e^{4i} b_i^2 g_i'' h_i''$. Since $b_0 = e$, $e^{4i} b_i^2$ is divided by e^{2i+2} for every $i \ge 0$, and we see that these g_{i+1}, h_{i+1} are the required. QED

A ring R is called a Hensel ring if R has only one maximal ideal, say M, and if it holds that: If $f(X)$, $g_0(X)$, $h_0(X)$ are monic polynomials in a variable X over R, $f - g_0 h_0 \in MR[X]$ and if $1 \in g_0 R[X] + h_0 R[X] + MR[X]$, then there are monic polynomials $g(X)$, $h(X)$ such that $g - g_0$, $h - h_0 \in MR[X]$, and $f = gh$.

The Hensel lemma proved above shows that a complete valuation ring is a Hensel ring. If a ring R is a homomorphic image of a Hensel ring, then R is obviously a Hensel ring.

THEOREM 5.11.7 If R is a normal ring with unique maximal ideal M, then the following four conditions are equivalent to each other:

(1) R is a Hensel ring.

(2) Every integral domain D which is integral over R has only one maximal ideal.

(3) If $f(X) = X^n + c_1 X^{n-1} + \cdots + c_n$ ($c_i \in R$) is such that $c_1 \notin M$, $c_2, \ldots, c_n \in M$, then $f(X)$ has a linear factor $X + a$ such that $a - c_1 \in M$.

(4) If $g(X) = X^n + d_1X^{n-1} + \cdots + d_n$ $(d_i \in R)$ is such that $d_n \in M$, $d_{n-1} \notin M$, then $g(X)$ has a linear factor $X + b$ with $b \in M$.

PROOF: (3) and (4) follow obviously from (1). Let us show first that each of (3) and (4) implies (2). Assume that D in (2) has maximal ideals P, Q $(P \neq Q)$. Take an element x of P outside of Q. Then $R[x]$ has at least two maximal ideals. Therefore we may assume that D is the least normal extension of R containing $R[x]$. Purely inseparable extensions do not increase the number of maximal ideals, and therefore we may assume that D is separable over R. Let G be the Galois group of D over R. The maximal ideals of D are P^σ $(\sigma \in G)$ (Theorem 4.7.7). Set $H = \{\sigma \in G \mid P^\sigma = P\}$, $S = \{x \in D \mid x^\sigma = x$ for all $\sigma \in H\}$, and $P' = P \cap S$. Prime ideals lying over P' are conjugate to P over S, and hence P is the unique prime ideal of D lying over P'. Thus, if $\sigma \notin H$, then $P^\sigma \cap S$ is a maximal ideal of S different from P'. Since $\#(G)$ is finite, D and S have only finite numbers of maximal ideals. Then there are elements a, b of S such that (i) $a \notin P'$ but a is in every other maximal ideal of S and (ii) $b \in P'$ but b is not in any other maximal ideal of S (Lemma 2.4.5). Let $f(X)$, $g(X)$ be minimal polynomials for a, b over R, and let $\sigma_1 = 1$, $\sigma_2, \ldots, \sigma_n$ be a set of representatives for $H\backslash G$. Since $a \in S$, $a_i = a^{\sigma_i}$ $(i = 1, \ldots, n)$ cover all the conjugates of a. Since $P^\sigma \neq P$ if $\sigma \notin H$, we see that, for $i \geq 2$, a_i is in P' and therefore $a_1, \ldots, a_n$ are exactly the conjugates of a. Thus elementary symmetric forms in the a_i of degree ≥ 2 belong to P, and hence also to M. Therefore $f(X)$ is of the form as in (3). (3) asserts that $f(X)$ is reducible, a contradiction. Similar observation on b gives us a contradiction in view of (4). Thus each of (3) and (4) implies (2). Now it suffices to prove that (2) implies (1). Assume that $f(X)$, $g_0(X)$, $h_0(X)$ are monic polynomials $(\in R[X])$ such that $f - g_0h_0 \in MR[X]$, $1 \in g_0R[X] + h_0R[X] + MR[X]$, and such that there are no $g,h \in R[X]$ such that $f = gh$ $(g - g_0,\ h - h_0 \in MR[X])$. If f is reducible, then the same holds for at least one factor of f, and therefore we may assume that f is irreducible. Since f is monic and since R is normal, it follows that f is irreducible over the field of fractions of R. Then $D = R[X]/fR[X]$ is an integral domain integral over R. $D/MD \cong (R/M)[X]/g^*h^*(R/M)[X]$ with $g^* = (g_0$ modulo $M)$, $h^* = (h_0$ modulo $M)$. Since $1 \in g_0R[X] + h_0R[X] + MR[X]$, we see that g^* and h^* have no proper common factor and we see that D has at least two maximal ideals, contradicting (2). QED

In general, if T is a normal extension of a normal ring R, P a maximal ideal of T, and if G is the Galois group, then $H = \{\sigma \in G \mid P^\sigma = P\}$ is called the splitting group of P and $S = \{x \in T \mid x^\sigma = x$ for every $\sigma \in H\}$ is called the splitting ring of P. Our proof of implication (3) or (4) $\Rightarrow$ (2) proves the following under the assumption that $\#(G)$ is finite: (i) $H \neq G$ if and only if S has at least two prime ideals lying over $P \cap R$ and (ii) there are elements a, b of S such that the minimal polynomials $f(X)$, $g(X)$ for a, b over R are of the form as in (3) and (4), respectively, with respect $M = P \cap R$, provided that $S \neq R$.

Assume that R is a normal ring with only one maximal ideal M. Let K be the field of fractions of R, K^* the separable algebraic closure of K, and R^* the integral closure of R in K^*. Take a maximal ideal P^* of R^*. Take the splitting ring T of P^*, and set $Q = P^* \cap T$, $S = T_Q$. Then S is unique within R-isomorphisms, because other maximal ideals are conjugate to P^* (Theorem 4.7.7). This S is called the henselization of R.

THEOREM 5.11.8 Under the notation as above, (1) S is a Hensel ring, and (2) if ψ is a homomorphism of R into a Hensel ring R' such that ψM is contained in the maximal ideal M' of R', then ψ is extended to a homomorphism ϕ of S into R' so that $\phi QS \subseteq M'$ and such an extension is unique.

PROOF: (1): Set $U = T - Q$. Then the integral closure of S in K^* is R^*_U. If this ring has a maximal ideal $P^{**}R^*_U$ (P^{**} being a maximal ideal of R^* different from P^*), then P^{**} is conjugate to P^* and lies over Q. Take an element a of P^* outside of P^{**}, and consider the least Galois extension R'' over T containing

a. Let H be the splitting group of $P^* \cap R''$ (over T), and let H^* be that of P^* (over R). $H^* \subseteq G(R^*/T)$. Let f be the natural mapping of $G(R^*/T)$ into $G(R''/T)$ (induced by the restriction on R''). Then $\sigma \in H^*$ implies that $f\sigma \in H$. This means that (the splitting ring of $P^* \cap R''$) $\subseteq$ (the splitting ring of P^*) = T, which is a contradiction.† Thus R^*_U has only one maximal ideal, and S is a Hensel ring.

(2): Let F be the set of normal rings R_λ having only one maximal ideal M_λ such that (i) $R \subseteq R_\lambda \subseteq S$, $QS \cap R_\lambda = M_\lambda$, (ii) $R_\lambda = (R^* \cap R_\lambda)_{P^* \cap R_\lambda}$, and (iii) there is a unique extension ψ_λ of ψ to a homomorphism of R_λ to R' such that $\psi_\lambda(M_\lambda) \subseteq M'$. $R \in F$, and F is not empty. Therefore we see easily that F is an inductive set and F has a maximal member, say T^*. Assume for a moment that $T^* \neq S$. Set $T' = T^* \cap R^*$, $P' = P^* \cap T'$, and $U' = T' - P'$. Then $T^* = T'_{U'}$, and therefore $R^*_{U'}$ is the integral closure of T^* in K^*. The splitting ring of $P^*R^*_{U'}$ is $T_{U'}$ and $S = (T_{U'})_{P^* \cap T_{U'}}$. Therefore, considering T^* instead of R, we may assume that $T^* = R$. By our proof of Theorem 5.11.7, there is a Galois extension R'' of R such that $\#(G(R''/R)) < \infty$ and R'' has maximal ideals other than $P^* \cap R''$. There is an element b of the splitting ring T' of $P^* \cap R''$ which is in $P^* \cap R''$ but not in any other maximal ideal, and b is a root of an irreducible monic polynomial $g(X) = X^n + c_1X^{n-1} + \cdots + c_n$ with $c_n \in M$, $c_{n-1} \notin M$, and $n > 1$. Since R' is a Hensel ring, $\psi(g(X))$ has a root $b^* \in M'$. If it has another root $b' \in M'$, then $\psi(g(X))$ is divisible by $(X - b^*)(X - b')$ and we have $\psi(c_{n-1}) \in M'$; a contradiction. Thus, such b^* is unique. $g(X)$ is irreducible over K because R is normal and g is monic. Therefore $R[b] = R[X]/gR[X]$, and ψ is extended to a homomorphism ϕ of $R[b]$ into R' so that $\phi b = b^*$; if we want to extend ψ so that $P^* \cap R[b]$ is mapped into M', then the extension must be ϕ, because of the unique-

---

† In Chap. 7, we shall study Galois theory of an infinite extension. There we see that $G(R^*/T) = H^*$. If we admit this result, the proof of this part is immediate.

ness of b^*. Now, ϕ is extended uniquely to a homomorphism of $R[b]_{P^*\cap R[b]}$ into R'. Since $R[b]_{P^*\cap R[b]}$ is normal as will be shown in the next theorem, we have a contradiction against the maximality of $T^* = R$. QED

THEOREM 5.11.9 Let R' be a Galois extension of a normal ring R with a finite Galois group, P' a maximal ideal of R', and T the splitting ring of P'. Set $P = P' \cap R$, $Q = P' \cap T$. Let $Q_1 = Q$, $Q_2, \ldots, Q_m$ be the maximal ideals of T lying over P. If $a \in Q$, $a \in Q_i$ for every $i \geq 2$, then $T_Q = R[a]_{Q\cap R[a]}$ and $T/Q = R/P$, $PT_Q = QT_Q$, $T_Q/QT_Q = R/P$.

PROOF: Considering R_P instead of R, we may assume that P is the unique maximal ideal of R. Set $I = \cap_{i\geq 2} Q_i$. Then $Q + I = T$. If $b' \in T$, then there is $b'' \in I$ such that $b' - b'' \in Q$. Thus in order to show that $T/Q = R/P$, it suffices to show that for every $b'' \in I$, there is an element b of R such that $b'' - b \in Q$. If $b'' \in Q$, then we may set $b = 0$. Assume that $b'' \notin Q$. Then by our proof of Theorem 5.11.7, the minimal polynomial $f(X) = X^n + c_1X^{n-1} + \cdots + c_n$ for b'' over R is such that $c_1 \notin P$, $c_2, \ldots, c_n \in P$. Then we have $b'' + c_1 \in PT \subseteq Q$. Thus we have proved that $T/Q = R/P$. Let us show next that $PT_Q = QT_Q$. If $z \in Q$ and if z is not in any other Q_i, then the minimal polynomial $g(X)$ for z over R is of the form $X^n + d_1X^{n-1} + \cdots + d_n$ with $d_n \in P$, $d_{n-1} \notin P$. Thus $z(z^{n-1} + d_1z^{n-2} + \cdots + d_{n-1}) \in d_nT$ and $z \in d_nT_Q \subseteq PT_Q$. If $y \in Q \cap I$, then $z + y$ satisfies the condition for z, and $z + y \in PT_Q$; hence $y \in PT_Q$. Thus, $QT_Q = (Q \cap I)T_Q \subseteq PT_Q$ and $QT_Q = PT_Q$. Now we consider $T' = R[a]_{Q\cap R[a]}$. Set $S = R[a] - (Q \cap R[a])$. Q is the unique maximal ideal of T containing a, and T_S is the integral closure of T' in the field L of fractions of T; therefore $T_S = T_Q$. If we take z as above, then our proof above shows that $z \in PR[z]_{Q\cap R[z]} \subseteq PT'[z]$, and therefore $T' + PT'[z] = T'[z]$, which implies that $T' = T'[z]$ by the lemma of Krull-Azumaya. Thus $z \in T'$. This (or our proof of Theorem 5.11.7) shows that T and T' share the same field of fractions. The derivative $g'(X)$ of the minimal polynomial $g(X)$ for z has a constant term not in P; hence

$g'(z) \notin Q$. If $x \in T$, then, since T is the derived normal ring of $R[z]$, we have $xg'(z) \in R[z]$ by Theorem 4.10.2, and therefore $x \in T'$. Therefore $T' = T_Q$. QED

THEOREM 5.11.10 If P is a prime ideal of a Hensel valuation ring R, then R_P is a Hensel valuation ring.

PROOF: Let L be an arbitrary algebraic extension of the field of fractions of R. Then the integral closure R' of R in L has only one maximal ideal, and hence R' is a valuation ring by Theorem 5.9.3. Set $U = R - P$. Since R' is a valuation ring, R'_U is also a valuation ring, and therefore R'_U has only one maximal ideal. Thus $R_U = R_P$ is a Hensel ring. QED

THEOREM 5.11.11 Let S^* be the henselization of a valuation ring R with maximal ideal P. Then S^* is a valuation ring with maximal ideal PS^*. Furthermore, $S^*/PS^* = R/P$ and every ideal of S^* is generated by elements of R. If v, v^* are valuations defined by R, S^* such that v^* is an extension of v, then the value groups of v, v^* coincide with each other.

PROOF: In view of Theorems 5.11.9 and 5.9.3 we have only to prove that if $y \in S^*$, then $yS^* = bS^*$ with $b \in R$, and therefore it suffices to prove:

THEOREM 5.11.12 Let R be a valuation ring of a field K, and let R' be a Galois extension of R with Galois group G of finite order. Let P' be a maximal ideal of R', H the splitting group of P', T' the splitting ring of P', and set $S' = T'_{Q'}$ with $Q' = P' \cap T'$. If $y \in T'$, then there is $b \in R$ such that $yS' = bS'$.

PROOF: For each prime ideal P'' of R' contained in P', we denote by $H(P'')$ the splitting group of P''. Then our proof of Theorem 5.11.10 shows that if $P^* \subseteq P''$ (prime ideals), then $H(P^*) \supseteq H(P'')$ $(H(0) = G)$. Let F be the set of all $H(P'')$, and let $G = H_0 \supset H_1 \supset \cdots \supset H_s = H$ be all of the members of F. For each i, let F_i be the set of prime ideals P'' $(\subseteq P')$ such that $H_i = H(P'')$. Then F_i is an inductive set and has a maximal member, say P_i $(P_s = P')$. We want to prove the assertion by induction on s.

(1) The case where $s = 1$: Take an element z of Q' which is

not in any other maximal ideal of T'. Then $S' = T'_{Q'} = R[z]_{Q' \cap R[z]}$ by Theorem 5.11.9, and the minimal polynomial $f(X) = X^n + c_1 X^{n-1} + \cdots + c_n$ for z is such that c_n is in the maximal ideal M of R and $c_{n-1} \notin M$ [$n = \#(G/H)$]; therefore $zS' = c_n S'$. Set $M_0 = P_0 \cap R$. If $\sigma \in G$, $\sigma \notin H$, then $z^\sigma \notin P'$ and therefore $z^\sigma - z \notin P'$. Therefore $f(X)$ modulo M_0 has no multiple root. If $f(X)$ modulo M_0 has a proper factor, then there must be at least two prime ideals of T' lying over M_0, which contradicts the fact that $H_0 = G$. Thus $f(X)$ modulo M_0 is irreducible, and therefore the ramification index and the degree of $S'' = R[z]_{P_0 \cap R[z]}$ over R_{M_0} are 1 and n, respectively. Thus every ideal of S'' is generated by elements of R. In particular, $yS'' = cS''$ with $c \in R$. Then y' which is either y/c or c/y is in S'. If y' is a unit in S', then $yS' = cS'$. Assume that $y' \in Q'S'$. Prime ideals of S' containing y' are not contained in any $T'_{M'}$ with maximal ideal $M' \neq Q'$. Therefore the approximation theorem implies that there is an element $y'' \in T'$ such that (i) $y''S' = y'S'$ and (ii) y'' is not in any maximal ideal M' ($\neq Q'$). Then y'' satisfies the condition for z above, and $y''S'$ is generated by an element of R. Thus $yS' = bS'$ with $b \in R$.

(2) Assume that $s > 1$. Let T'' be the splitting ring of P_1, and set $M_1 = P_1 \cap R$, $Q'' = P_1 \cap T''$. By induction on s applied to $T''_{Q''}$ over R_{M_1}, we see that every ideal of $T''_{Q''}$ is generated by elements of R: The case where $s = 1$ applied to S' over $T''_{P' \cap T''}$ implies that every ideal of S' is generated by elements of $T''_{P' \cap T''}$. In particular, $yS' = dS'$ with $d \in T''$, and $dT''_{Q''} = cT''_{Q''}$ with $c \in R$. Then y' which is either y/c or c/y is in S'. If y' is a unit in S', then $yS' = cS'$, and we assume that $y' \in Q'S'$. Then every prime ideal of S' containing y' is not contained in any $T'_{M'}$ (M' being a maximal ideal $\neq Q'$), and we prove the case as in the last step of (1). QED

In the case of a valuation of rank 1, there is an interesting relationship between the henselization and the completion, namely:

THEOREM 5.11.13 Let R be the valuation ring of a nonarchimedean valuation v, and R^* the completion of R. Then the set R'' of separably algebraic elements in R^* over R forms the henselization of R.

PROOF: By Theorem 5.11.8, the natural injection $R \to R^*$ is extended to a homomorphism ψ of the henselization S into R^*. $(\psi^{-1}(0)) \cap R = 0$ and $\psi^{-1}(0) = 0$ by Theorem 5.11.11. Thus ψ is an injection, and we regard S as a subring of R^*. By our definition of a henselization, elements of S are separably algebraic over R and $S \subseteq R''$. Let L be the field of fractions of R''. Then $R'' = R^* \cap L$, and therefore R'' is a valuation ring of L (Corollary 5.7.3) and defines the restriction w of the completion v^* of v. Let R' be the integral closure of S in L. Since S is a Hensel ring, R' has only one maximal ideal, and therefore R' is a valuation ring. Thus $R'' = R'$ (Theorem 5.7.2). Let y be an arbitrary element of R', and let $f(X) = X^n + c_1X^{n-1} + \cdots + c_n$ be the minimal polynomial for y over S. Assume that $n > 1$, let $y_1 = y$, y_2, $\ldots$, y_n be the roots of $f(X)$, and let d be the discriminant of $f(X)$ $(d = \Pi_{i<j}(y_i - y_j)^2)$. Since y is separable, $d \neq 0$, and there is an element $(0 \neq)$ $b \in R$ such that $wb < wd$. Since $y \in R^*$, there is an element $c \in R$ such that $y - c \in bR^*$. Then, considering $y - c$ instead of y (note that d is unchanged by this replacement), we may assume that $wy \leq wb < wd$. Since S is a Hensel ring, the extension to L of the valuation defined by S must be unique. This means that $w(y_i) = w(y_j)$ for any i, j. Then $wb < wd = \Pi_{i<j} w(y_i - y_j)^2 \leq wy^2 < wb$, which is a contradiction. Thus $n = 1$ and $S = R'$. QED

Using this result, we can obtain a form of the Hensel lemma as follows:

THEOREM 5.11.14 Let R be a Hensel valuation ring with maximal ideal M. Let $f(X)$, $g_0(X)$, $h_0(X)$ be polynomials in X over R, and let $(0 \neq)$ $d \in R$. Let P be M or $\sqrt{dR}$ according to whether d is a unit or a nonunit. Assume that (1) g_0 is monic, (2) $f - g_0h_0 \in d^2PR[X]$, and (3) $d \in g_0R[X] + h_0R[X] + dPR[X]$; then there are $g(X)$, $h(X) \in R[X]$ such that (1') g is monic, (2') $f = gh$, and (3') $g - g_0$, $h - h_0 \in dPR[X]$.

PROOF: (i) The case where Krull dim $R = r < \infty$: If $r = 0$, the assertion is obvious. Assume that $r = 1$. Theorem 5.11.4 implies that there are such g, h with coefficients in the completion R^* of R. By Theorem 5.11.12, it suffices to show that the coefficients of g, h are separably algebraic over R. Let $b_1, \ldots, b_m$ and $c_1, \ldots, c_n$ be roots of g and h, respectively. Since g is monic, $g = \Pi(X - b_i)$, $h = c\Pi(X - c_j)$, where c is the coefficient of the highest degree term in f. Therefore the coefficients of g, h are algebraic over R; if some of the coefficients are inseparable over R, then f has a multiple root, say a, which is b_i and c_j for some i, j. Then g and h have a common root and hence have a proper common factor k $(k(a) = 0)$. Since $d \in g_0R^*[X] + h_0R^*[X] + dPR^*[X] = gR^*[X] + hR^*[X] + dPR^*[X]$, $d = kk' + dqk''$ with $k',k'' \in R^*[X]$, $q \in P$. Then, setting $X = a$, we have $d = dqk''(a)$. Since $q \in M$ in any case, we have $d = 0$; a contradiction. Thus coefficients of g, h are separable and $g,h \in R[X]$. Assume now that $r > 1$. Let Q be a prime ideal of height 1. The R/Q is a Hensel ring and hence is a Hensel valuation ring (Corollary 5.7.4). R_Q is also a Hensel valuation ring by Theorem 5.11.10.

(I) The case where $d \in Q$: $P = Q$ in this case, and hence by the case where $r = 1$ applied to R_Q, we see that there are $g,h \in R_Q[X]$ such that $f = gh$, g monic and $g - g_0$, $h - h_0 \in dQR_Q[X]$. Since $QR_Q = Q$ (Corollary 5.7.4), we see that these g, h are the required polynomials. (II) The case where $d \notin Q$: $P \neq Q$ in this case and an induction applied to R/Q implies that there are $g_1,h_1 \in R[X]$ such that $f - g_1h_1 \in QR[X]$, g_1 monic and $g_1 - g_0$, $h_1 - h_0 \in dPR[X]$. Let d_1 be the resultant of g_1, h_1. Since (g_1 modulo Q) and (h_1 modulo Q) have no common root, we have $d_1 \notin Q$ and $d_1 \in g_1R[X] + h_1R[X]$. Then we apply this to R_Q, and we see the existence of g, h as in case (I).

(ii) The general case: Consider the relations $f - g_0h_0 = d^2qk$ $(q \in P,\ k \in R[X])$, $d = g_0k' + h_0k'' + dq'k^*$ $(q' \in P;\ k',k'',k^* \in R[X])$. Let L be the subfield generated by d, q, q' and coefficients of f, g_0, h_0, k, k', k'', k^*, and let R' be the henselization of $R \cap L$. Then R' can be identified with a subring of R (Theorem 5.11.8). Case (i) implies the existence of $g, h \in R'[X] \subseteq R[X]$. QED

We add here an interesting application of the theorem above and the approximation theorem:

THEOREM 5.11.15 Let R be a Hensel valuation ring of a field K such that (i) $R \neq K$ and (ii) for every prime ideal P different from the maximal ideal, the field L_P of fractions of R/P has a proper extension field E_P which is separably algebraic over L_P. If R' is a Hensel valuation ring of K, then either R' contains R or R contains R'.

PROOF: Assume the contrary, and set $D = R \cap R'$. Let M, M' be maximal ideals of R, R', respectively, and let Q be the maximal member in the set of common prime ideals of R and R'. By our assumption (ii), there is an element $a \in E_Q$ ($a \notin L_Q$) which is integral over R/Q. Let $f(X) = X^n + b_1X^{n-1} + \cdots + b_n$ be the minimal polynomial for a over R/Q ($b_i \in R/Q$), and set $g(X) = X^n + c_1X^{n-1} + \cdots + c_n$ with $c_i \in R$ such that (c_i modulo Q) $= b_i$. Since $f(X)$ is irreducible, $g(X)$ must be irreducible. Since $f(X)$ is separable, the discriminant d_g of $g(X)$ is not in Q. On the other hand, let $y_1, \ldots, y_n$ be elements of D such that $y_i - y_j \notin Q$ for any i, j, and set $h(X) = \Pi(X - y_i)$. Write $h(X) = X^n + e_1X^{n-1} + \cdots + e_n$, and let d_h be the discriminant of $h(X)$. Since $d_g, d_h \notin Q$, there are $z_i \in K$ such that $z_i - c_i \in d_g{}^2M\sqrt{d_g R}$, $z_i - e_i \in d_h{}^2M'\sqrt{d_h R}$ ($i = 1, \ldots, n$) (approximation theorem). Consider $k(X) = X^n + z_1X^{n-1} + \cdots + z_n$. Since discriminants of monic polynomials of a fixed degree are expressed in a definite polynomial of coefficients of them, we see as for the discriminant d_k of $k(X)$ that (1) $d_k - d_g \in d_g{}^2M\sqrt{d_g R}$ because $z_i - c_i \in d_g{}^2M\sqrt{d_g R}$ and (2) $d_k - d_h \in d_n{}^2M'\sqrt{d_h R}$ similarly. Set $h_1 = X - y_1$, $h_2 = h/h_1$. Then the resultant of h_1, h_2 is a factor of d_h. Therefore $k(X)$ is reducible by Theorem 5.11.14, and the factorization is applied to $g(X)$. Then we see that $g(X)$ is reducible, a contradiction. QED

5.12 THEOREM OF LÜROTH, CONTINUED

THEOREM 5.12.1 Let x be a transcendental element over a field K. Let V_x be a weakly discrete valuation ring of $K(x)$, namely, V_x has prime ideals $P_0 = 0 \subset P_1 \subset \cdots \subset P_m$ = (the maximal ideal M_x) such that every $(R_{P_i})/P_{i-1}$ is discrete $(i = 1, \ldots, m)$. Set $V = V_x \cap K$, and let M be the maximal ideal of V. Assume that $K^* = V_x/M_x$ is not algebraic over $K' = V/M$. Then there are a field K'' and an element t such that (i) $K' \subseteq K'' \subset K^*$, (ii) t is transcendental over K'', and (iii) $K^* = K''(t)$. K'' is algebraic over K'.

PROOF: If $m = 0$, then $K' = K$, $K^* = K(x)$, and the assertion is obvious. Assume that $m = 1$. Then V_x is discrete, and therefore V is also discrete and $M = pV$ with $p \in M$. If $p = 0$, then K^* must be algebraic over K (cf. Exercise 5.2.1). Therefore $p \neq 0$. Let v be an additive valuation defined by V_x. If $vx < 0$, $v(x^{-1}) > 0$, and therefore we may assume that $vx \geq 0$. Set $R_1 = V[x]_{M_x \cap V[x]}$. Then either (1) the maximal ideal M_1 of R_1 is generated by two elements and is of height 2 or (2) R_1 is a discrete valuation ring; this is so because $M_x \cap V[x]$ contains p and $V[x]/pV[x] \cong K'[x]$; therefore if $M_x \cap V[x] \neq pV[x]$, then $M_x \cap V[x]$ is generated by two elements and $\mathrm{ht}\,(M_x \cap V[x]) = 2$; on the other hand, if $M_x \cap V[x] = pV[x]$, then R_1 is a discrete valuation ring with maximal ideal pR_1. Starting with R_1, we construct R_i inductively as follows: Assume that R_s is constructed. If R_s is a valuation ring, then let R_s be the last term. In the other case, the maximal ideal M_s of R_s is assumed to be generated by two prime elements, say y_s and z_s, and $\mathrm{ht}\, M_s = 2$. We may assume that $v(y_s) \leq v(z_s)$. Then we set $R_{s+1} = R_s[z_s/y_s]_{M_x \cap R_s[z_s/y_s]}$. Since $R_s[z_s/y_s] \cong R_s[X]/(y_sX - z_s)R_s[X]$, y_s is a prime element in $R_s[z_s/y_s]$ and $R_s[z_s/y_s]/y_sR_s[z_s/y_s] \cong (R_s/M_s)[X]$. Thus R_{s+1} enjoys one of (1) and (2) for R_1. We want to show that this sequence terminates. Indeed, assume that $\mathrm{ht}\, M_i = 2$. Then R_i/M_i is algebraic over K' as is easily seen by induction on i. By our assumption, K^* contains a transcendental element, say (u modulo M_x), over K'. Write $u = f_i/g_i$ ($f_i, g_i \in R_i$). If $g_i \in M_i$, then $vu = 0$ and $v(f_i) = v(g_i) > 0$, $f_i \in M_i$; and f_i/y_i, $g_i/y_i \in R_{i+1}$ and we may choose g_{i+1} to be g_i/y_i. Then $v(g_{i+1}) < v(g_i)$. Since v is discrete, this sequence must terminate. Thus, at some step, R_i is a discrete valuation ring of K(x) and $R_i \subseteq V_x$, and therefore $R_i = V_x$. Take such R_i, and compare R_i with R_{i-1}. Let K" be R_{i-1}/M_{i-1}. Since $K^* = R_i/M_i$ is generated by the class t of z_{i-1}/y_{i-1}, we see that these K" and t are the required.

Now we assume that $m > 1$. $V'_x = V_x/P_1$ is not algebraic over $V' = V/Q_1$ (here $Q_1 = P_1 \cap V$) in view of Theorem 5.9.3. Therefore by the case where $m = 1$ applied to $(V_x)_{P_1}$ over V_{Q_1}, we see that there is a transcendental element t' over V' such that $L^* = (V_x)_{P_1}/P_1$ is generated by t' over a field L which is containing and algebraic over V'. Then our induction applied to V'_x over $V'_x \cap L$ proves the assertion. QED

Now we prove a result we needed in Sec. 4.12.

THEOREM 5.12.2 Assume that K is a subfield of a field L and that x is a transcendental element over L. If x, t ($t \in L(x)$) are algebraically independent over K, and if $L(x) = K(x,t)$, then $L \cong K(t)$.

PROOF: Let v be the x-adic additive valuation with respect to L[x] such that $vx = 1$. Then the valuation ring V_x of v must be $L[x]_{xL[x]}$. Set $L^* = K(t)$. If v is trivial on L^*, then $V_x = L^*[x]_{xL^*[x]}$ and $L \cong V_x/xV_x \cong L^*$. Assume that v is nontrivial on L^*. Set $V = V_x \cap L^*$, $M = xV_x \cap V$. Since v is trivial on K, we see that V/M is algebraic over K. Thus $L = V_x/xV_x$ is not algebraic over V/M. Thus we can apply Theorem 5.12.1, and we see that there is a field K" such that $K \subseteq V/M \subseteq K'' \subset L$ and L is generated by a transcendental element over K". Since $L(x) = K(x,t)$, K is algebraically closed in L(x), and hence in L also. Therefore $K = K''$. QED

EXERCISE 5.1

1. In the field $K(X)$ (K a field and X transcendental over K), for $0 \neq f = g(X)/h(X)$ ($g,h \in K[X]$), we define $\deg f = \deg g - \deg h$. Let r be a positive real number and define v by $vf = r^{\deg f}$, $v0 = 0$. Discuss in what case this v is a multiplicative valuation.
2. Prove that a field which is algebraic over a finite field has no nontrivial valuation.

EXERCISE 5.2

1. Consider $K(X)$ with a finite field K and a transcendental element X. Prove that if v is a nontrivial valuation of $K(X)$, then v is an f-adic valuation with respect to (i) $K[X]$ in case $v(X) \geq 0$, with a monic irreducible polynomial f, or (ii) $K[X^{-1}]$ in case $v(X) < 0$, with $f = X^{-1}$.

   Prove the same without assuming that K is a finite field but assuming that the restriction of v on K is trivial.
2. Let L be an algebraic extension field of a field K, and let v be a multiplicative valuation of L. Prove that if the restriction w of v on K is trivial, then v is trivial along the following line:

   (i) Prove that v is nonarchimedean. (ii) Prove that the valuation ring R of v is normal. (iii) Then, using (ii), prove that $R = L$.

EXERCISE 5.3

1. A set B of open sets in a topological space S is called a <u>basis</u> of open sets if every open set is the union of certain members of B.
   (1) Prove that if B_0 is a subbasis of open sets in S, then the set B, of sets which are intersections of finite numbers of members of B_0, forms a basis of open sets.
   (2) Prove that a set W of open sets of S forms a basis

of open sets if and only if it holds that if U is an open set and if $P \in U$, then there is $V \in W$ such that $P \in V \subseteq U$.

2. Let S be a metric space, let S' be a dense subset of S, and let $r_1, r_2, \ldots$ be a sequence of positive numbers which tends to 0. Prove that $B = \{r_i\text{-neighborhood of } P \mid P \in S';\ i = 1, 2, \ldots\}$ forms a basis of open sets.
3. Let P be a proposition expressed by open sets, closed sets, $\supseteq$, $=$, $\cup$, $\cap$. The dual of P is the new proposition P^* obtained from P by interchanging (i) open set and closed set (ii) $\supseteq$ and $\subseteq$, and (iii) $\cup$ and $\cap$ (= is unchanged). [For instance, (1) and (2) in Theorem 5.3.7 are dual to each other.] Prove that P is equivalent to P^*.
4. Let $f : S \to S'$ and $g : S' \to S''$ be continuous mappings of topological spaces. Prove that $gf : S \to S''$ is also continuous.

EXERCISE 5.4

1. Confirm that H in each the case below is a topological group:
   (1) H is the direct product of topological groups $G_1, \ldots, G_n$, namely, $H = G_1 \times \cdots \times G_n$ as a group and as a topological space simultaneously.
   (2) H is the additive group of rational integers in which the set of all $U_n(m) = \{x \in H \mid m - x \text{ is a multiple of } n\}$ ($m \in H$; n natural number) is a subbasis of open sets.
   (3) H is the additive group of rational integers in which a natural number $N > 1$ is fixed and the set of all $U_n(m) = \{x \in H \mid m - x \text{ is divisible by } N^n\}$ ($m \in H$; n natural number) is a subbasis of open sets.
   (4) H is a subgroup of a topological group G with induced topology from G.
   (5) Let G' be a normal subgroup of a topological group G. If G' is a closed set, we define the residue class group $H = G/G'$ not just as a group but with topology defined by taking $\{fU \mid U \text{ open in } G\}$ (f is the natural mapping of G to H) as a subbasis of open sets.

2. In (5) above, prove that if G' is not a closed set, then under the topology defined similarly, H becomes a topological space which is not a T_0-space.
3. In (5) above, show by an example that even if F is a closed set in G, fF need not be a closed set of H.
4. Let U be an open set in a topological group G. Prove that if $S \subseteq G$, then $SU = \{su \mid s \in S, u \in U\})$ is an open set of G.

EXERCISE 5.5

1. Let R be the valuation ring of a nonarchimedean valuation v. Construct the completion R^* of R adapting the construction of the completion of a field. Then prove that if $0 \neq a \in R$, then $R^*/aR^* \cong R/aR$.
2. The formal power series ring in $X_1, \ldots, X_n$ over a ring R is the set F of formal sums $\sum c_{(i)}X^{(i)}$ [here, $(i) = (i_1,\ldots,i_n)$ with nonnegative rational integers i_j; $X^{(i)} = X_1^{i_1}\cdots X_n^{i_n}$; $c_{(i)} \in R$]. In F, addition and multiplication are defined by

$$\sum c_{(i)}X^{(i)} + \sum d_{(i)}X^{(i)} = \sum (c_{(i)} + d_{(i)})X^{(i)}$$

$$(\sum c_{(i)}X^{(i)})(\sum d_{(i)}X^{(i)}) = \sum (\sum_{(i)+(j)=(k)} c_{(i)}d_{(j)})X^{(k)}$$

Confirm that F is a ring. This F is denoted by $R[[X_1,\ldots,X_n]]$. Each element of F is called a formal power series.

Prove that if R is a field and if $n = 1$, then F is a discrete valuation ring.
3. Let R be a complete discrete valuation ring with maximal ideal qR. Assume that $K = R/qR$ is a perfect field of characteristic $p \neq 0$. $a, a_i, b, \ldots$ will denote elements of R; adding $'$ will mean their residue classes modulo qR.

(I) Let a' $(\neq 0)$ be an element of K, and let a_n be such that $a_n'^{p^n} = a'$. Prove that the sequence $\{a_n^{p^n}\}$ is a Cauchy sequence. Show that $\psi a' = \lim a_n^{p^n}$ gives an injection ψ of the multiplicative group $K - \{0\}$ into R. Set $K^* = \{\psi a' \mid 0 \neq a' \in K\}$. Prove that every element of R is expressed uniquely in the form $\sum a_i q^i$ $(a_i \in K^* \cup \{0\})$ [$\sum a_i q^i$ means $\lim_n (\sum_{i<n} a_i q^i)$].

(II) Assume furthermore that R itself is of characteristic 0 and $q = p$. Each element $a = \sum a_i p^i$ $(a_i \in K^* \cup \{0\})$ of R may be represented by a vector $(a_1, a_2, \ldots)$. Then such a vector is called a <u>Witt vector</u> of infinite length over K. Each element of R/p^nR can be represented by $(a_0, a_1, \ldots, a_{n-1})$ $(a_i \in K^* \cup \{0\})$ similarly. Such a vector is called a <u>Witt vector</u> of length n over K. R and R/p^nR as sets of these vectors are called <u>Witt rings</u> of infinite length and of length n, respectively. We dare ask the following, though they are too difficult to be asked here.

Prove that for each $m = 1, 2, \ldots$, there are polynomials $f_m(X_0, X_1, \ldots, X_m, Y_0, \ldots, Y_m)$ and $g_m(X_0, \ldots, X_m, Y_0, \ldots, Y_m)$ with coefficients in the prime field such that if $a = (a_0, \ldots)$, $b = (b_0, \ldots)$ are Witt vectors of the same length, then

$$a + b = (a_0 + b_0, f_1(a_0, a_1, b_0, b_1), \ldots, f_m(a_0, \ldots, a_m, b_0, \ldots, b_m), \ldots)$$
$$ab = (a_0 b_0, g_1(a_0, a_1, b_0, b_1), \ldots, g_m(a_0, \ldots, a_m, b_0, \ldots, b_m), \ldots)$$

Prove also that $pa = (0, a_0^p, \ldots, a_{m-1}^p, \ldots)$. Prove on the other hand that the mapping F such that $a^F = (a_0^p, a_1^p, \ldots, a_m^p, \ldots)$ gives an automorphism of the Witt rings over K. This F is called the <u>Frobenius map</u>.

4. Let K be a complete field with respect to a valuation v. Let M be a K-module with linearly independent base $u_1, \ldots, u_n$, and a topology so that (i) M is a metric space and forms a topological group under addition and (ii) denoting by $|m|$ the distance between m and 0, it holds that $|am| = (va)|m|$ for any $a \in K$, $m \in M$. Prove that the topology on M is uniquely determined by these conditions and coincides with the one as follows: Let each Ku_i be naturally homeomorphic to K, and let M be the product space of $Ku_1, \ldots, Ku_n$.

EXERCISE 5.6

1. Let t be transcendental over the complex number field $\mathbb{C}$. Prove that $\mathbb{C}(t)$ has an archimedean valuation v and that its restriction on $\mathbb{C}$ cannot be equivalent to the absolute value.

2. Characterize fields which have no archimedean valuations.
3. Let K be a complete field with respect to a valuation v, and let L be an algebraic extension of finite degree over K. Prove the uniqueness of an extension of v to L by using Exercise 5.5.4.

EXERCISE 5.7

1. Prove that a valuation ring R is not noetherian if Krull dim $R \geq 2$.
2. Let R_λ $(\lambda \in \Lambda)$ be valuation rings contained in a field K. Prove that $D = \cap R_\lambda$ is a normal ring.
3. If an additive group G has a subset P satisfying the following conditions (i) to (iii), then there is one and only one order so that G is an ordered group in which $a > 0$ if and only if $a \in P$.
   (i) $a,b \in P$ implies $a + b \in P$.
   (ii) $0 \neq a \notin P$ implies $-a \in P$.
   (iii) $0 \notin P$.
4. Let R be a valuation ring of a field K, and let S be a subring of an extension field L of K. Prove that if $S \supseteq R$ and if S has a prime ideal P lying over the maximal ideal M of R, then $S \cap K = R$. Prove also that if furthermore S is a valuation ring, then a valuation v defined by R is extended to a valuation w of L whose ring is S, and in this case the rank of w is not less than the rank of v.
5. Let R be the valuation ring of an additive valuation v of a field K, and let P be a prime ideal of R. Set $G_P = \{va \mid a \in K, a \notin P, a^{-1} \notin P\}$. Prove that (1) G_P is an additive subgroup of the value group G of v; (2) if $0 < g \in G$, $g \notin G_P$, then for any natural number n and any $h \in G_P$, it holds that $nh < g$; and (3) G/G_P is naturally isomorphic to the value group of the valuation defined by R_P.
6. With the natation as above, a subgroup of G satisfying Condition

(2) is called an isolated subgroup of G. Prove that every isolated subgroup of G is obtained as G_P with suitable P.

7. Prove that (1) a valuation ring is noetherian if and only if it is a discrete valuation ring (for the difinition, see Theorem 5.9.7) and (2) an additive valuation v of rank n is weakly discrete (for the difinition, see Theorem 5.12.1) if and only if its value group is isomorphic to the n-ple product of the additive group of rational integers (the order being the lexicographical order).

8. Let R be the valuation ring of an additive valuation v of rank n, and let $0 = P_0 \subset P_1 \subset \cdots \subset P_n$ be all of the prime ideals of R. Prove (1) that the value group of v is isomorphic to a subgroup of the n-ple product of the additive group $\mathbb{R}$ of real numbers (the order being the lexicographical order). Let G_i be the value group of a valuation w_i defined by R_{P_i}/P_{i-1}. Then prove (2) that if every G_i is divisible (i.e., it holds that if $g \in G_i$ and if m is a natural number, then there is $h \in G_i$ such that $mh = g$), then the value group G of v is naturally isomorphic to the direct sum $G_1 \oplus \cdots \oplus G_n$ (the order being the lexicographical order).

9. Under the notation as above, but without assuming the divisibility of G_i, each G_i or G is a module over the ring $\mathbb{Z}$ of rational integers. In that sense, we can speak of a maximal linearly independent subset B_i, B of G_i, G, respectively. $\#(B_i)$, $\#(B)$ are called the rational ranks of G_i, G. Prove that $\#(B) = \sum \#(B_i)$. Assume next that R contains a field F. Prove that $\text{trans.deg}_F\, R \geq \#(B) + \text{trans.deg}_F\, R/P_n$.

EXERCISE 5.8

1. Let $v_1, \ldots, v_n$ be additive valuations of a field K and let $R_1, \ldots, R_n$ be their valuation rings. Find a necessary and sufficient condition on R_i for the validity of the following: If $a_1, \ldots, a_n, b_1, \ldots, b_n \in K$ and if $b_1 \cdots b_n \neq 0$, then there is $c \in K$ such that $v_i(c - a_i) > v_i(b_i)$ for $i = 1, \ldots, n$.

2. Let $R_1, \ldots, R_n$ be valuation rings of a field K. Assume that a prime ideal P of R_1 is not contained in any other R_i. Prove that there is an element b of P which is a unit in every R_i $(i \geq 2)$.

EXERCISE 5.9

1. Let K be a field of characteristic $p \neq 0$. Prove that there is a discrete valuation ring R such that (i) R is of characteristic 0, (ii) the maximal ideal of R is pR, (iii) R is complete, and (iv) $R/pR \cong K$.
2. Under the notation of Theorem 5.9.6, prove that $\sum e_i f_i = [L : K]$ if there are a prime ideal Q of R and an element a of L such that (i) $L = K(a)$, (ii) a is integral over R and the discriminant d_f of the minimal polynomial $f(X)$ for a over R is not in Q, and (iii) R/Q is discrete.

EXERCISE 5.10

1. Let L be an algebraic extension field of finite degree of either the rational number field Q or the field $K(X)$ with a finite field K and a transcendental element X. Assume that the product formula holds for L with respect to two sets V and V' of valuations. Prove that there is a positive number t such that $V' = \{v^t \mid v \in V\}$.

EXERCISE 5.11

1. Prove that if a valuation ring R has a prime ideal P so that both R/P and R_P are Hensel rings, then R is a Hensel ring.
2. Assume that R is a Hensel integral domain. Prove that if T is a normal ring integral over R, then T is a Hensel ring.
3. Prove that the formal power series ring $K[[X_1,\ldots,X_n]]$ over a field K is a Hensel ring.

4. We denote by $K((X))$ the field of fractions of $K[[X]]$ (K being a field). Prove that if R is a Hensel valuation ring of $K((X))$ of Krull dimension 1, then $R = K[[X]]$. Find other Hensel valuation rings of $K((X))$.

EXERCISE 5.12

1. Assume that K, k' are finitely generated fields over a field k and that $t, x_1, \ldots, x_n$ and $x_1, \ldots, x_n$ are algebraically independent over k', K, respectively. Prove that there are a field k'' containing k and a transcendental element z over k'' such that $K = k''(z)$ if it holds that $K(x_1,\ldots,x_n) = k'(t, x_1,\ldots,x_n)$.

CHAPTER 6

ORDERED FIELDS

6.1 ORDERED FIELDS AND FORMALLY REAL FIELDS

An ordered field K is a field which is an ordered set satisfying: (1) $a \geq b$, $c \geq d$ imply $a + c \geq b + d$, (2) $a > b$ implies $-a < -b$, and (3) $a > 0$, $b > 0$ implies $ab > 0$.

If K is of characteristic $p \neq 0$ and if $a > 0$, then $0 > -a = (p - 1)a = a + \cdots + a > 0$, which is a contradiction. Thus:

THEOREM 6.1.1 Every ordered field is of characteristic zero.

An injection (or an isomorphism) ψ of an ordered field K to an ordered field K' is an injection (or an isomorphism, respectively) of a field such that it enjoys the property that $a > b$ if and only if $\psi a > \psi b$; ψ may be called an order-isomorphism. We say that K is order-isomorphic to K', or that K is isomorphic to K' as ordered fields if there is such an isomorphism.

A formally real field K is a field in which $X_1^2 + X_2^2 + \cdots + X_n^2 = 0$ has no nontrivial solution for any natural number n. It follows immediately that a formally real field is of characteristic zero. Furthermore, it really holds:

THEOREM 6.1.2 An ordered field is formally real. Conversely, if K is a formally real field, then we can make K into an ordered field (by giving a suitable order).

The first half of the theorem is immediate from the following:

LEMMA 6.1.3 If K is an ordered field, then $a^2 \geq 0$ for any $a \in K$.

PROOF: If $a > 0$, then $a^2 \geq 0$; if $a < 0$, then $-a > 0$ and $a^2 > 0$. QED

The last half of Theorem 6.1.2 will be proved after Theorem 6.1.8.

A <u>real closed</u> field K is a formally real field such that no proper algebraic extension of K is formally real. When a field L is given, the set F of formally real fields contained in L forms an inductive set unless F is empty. Therefore we see:

THEOREM 6.1.4 If K is a formally real field, then there is a real closed field K^* ($\supseteq K$) which is algebraic over K.

LEMMA 6.1.5 Let K be a field and set $S = \{a_1^2 + \cdots + a_n^2 \mid a_i \in K,\ n$ a natural number$\}$. Then it holds that (i) if $a,b \in S$ then $a+b, ab \in S$. (ii) If $0 \neq a \in S$ then $a^{-1} \in S$. Furthermore, $S = K$, unless K is formally real or the characteristic of K is 2.

PROOF: (i) is immediate. As for (ii), $a = a_1^2 + \cdots + a_n^2$ and $a^{-1} = (a_1/a)^2 + \cdots + (a_n/a)^2 \in S$. Assume that K is not formally real. Then $-1 = a_1^2 + \cdots + a_n^2$ $(a_i \in K)$. Then, for arbitrary $b \in K$, we have $4b = (1 + b)^2 + (a_1^2 + \cdots + a_n^2)(1 - b)^2$. Therefore $b \in S$ unless the characteristic of K is 2. QED

THEOREM 6.1.6 If K is a real closed field and if $a \in K$, then there is $b \in K$ such that $a = \pm b^2$.

PROOF: Assume that $a \neq b^2$ for every $b \in K$. Then $K(\sqrt{a})$ is a proper algebraic extension of K and is not formally real. Thus $-1 = \sum_{i=1}^n (b_i + c_i\sqrt{a})^2$ $(b_i, c_i \in K)$. Then $1 + \sum b_i^2 + a(\sum c_i^2) = 0$, $\sum b_i c_i = 0$. Then by S in Lemma 6.1.5, we have $-a \in S$. If $a \in S$, we have $-1 \in S$, which contradicts that K is formally real. Therefore $a \notin S$. If $-a \neq b^2$ for every $b \in K$, we have $a \in S$ similary. Therefore $-a = b^2$ for some $b \in K$. QED

LEMMA 6.1.7 Assume that a subset P of a field K enjoys the following properties (1) to (3), then K becomes an ordered field by defining $a > b$ if and only if $a - b \in P$. (1) $a,b \in P$ implies $a + b$, $ab \in P$. (2) $0 \notin P$. (3) If $0 \neq a \in K$, then a or $-a \in P$.

PROOF: Set $N = \{-a \mid a \in P\}$. If $a \in P \cap N$, then $0 = -a + a \in P$, which contradicts (2). Thus $P \cap N$ is empty and therefore, if $a,b \in K$, then it holds one and only one of $a > b$, $a = b$, $a < b$. If $a \geq b$, $b \geq c$, then $a - b$, $b - c \in P \cup \{0\}$, and $a - c \in P \cup \{0\}$. Therefore K is linearly ordered. It is easy to see the compatibility of the order with addition and multiplication. QED

THEOREM 6.1.8 A real closed field K can be made into an ordered field, and such an order is unique. Any automorphism of K is an order-isomorphism.

PROOF: Set $P = \{a^2 \mid 0 \neq a \in K\}$. Theorem 6.1.6 shows that P enjoys the conditions in Lemma 6.1.7. By lemma 6.1.3, every element of P must be positive, and therefore the uniqueness is proved. Under any automorphism, P is mapped to P, which proves the last assertion. QED

PROOF of the last half of THEOREM 6.1.2: If K is a formally real field, then there is a real closed field L containing K (Theorem 6.1.4), and L is made into an ordered field (Theorem 6.1.8). QED

THEOREM 6.1.9 An ordered field K is real closed if and only if K enjoys the conditions (1) if $0 < a \in K$, then there is $b \in K$ such that $a = b^2$; and (2) if $f(X)$ is a monic polynomial in X of odd degree over K, then there is at least one root of $f(X)$ in K. In this case, $K(\sqrt{-1})$ is algebraically closed.

PROOF: The last statement is proved by an adaption of our proof of Theorem 3.11.4. Then there is no proper algebraic extension of K which is formally real. This proves the if part. Assume conversely that K is real closed. Then (1) follows from Theorem 6.1.6. As for (2), we use an induction on $n = \deg f(X)$. If $f(X)$ is reducible, then some factor is of odd degree, and this case is proved by induction. So we assume that $f(X)$ is irreducible and $n > 1$. Let b be a root of $f(X)$. Since $K(b) \neq K$, $K(b)$ is not formally real, and $-1 = d_1^2 + \cdots + d_m^2$ $[d_i \in K(b)]$. Each d_i is expressed as $g_i(b)$ with $g_i(X) \in K[X]$, $\deg g_i < n$. Then $1 + g_1(X)^2 + \cdots +$

$g_m(X)^2$ has b as a root and hence is divisible by $f(X)$. Let the quotient be $h(X)$. Then $\deg h(X) \leq 2(n - 1) - n = n - 2$. $\deg h(X)$ is odd because $\deg f(X)$ is odd and $\deg (1 + \sum g_i(X)^2)$ is even. Therefore, by our induction hypothesis, there is a root c of $h(X)$ in K. Then $1 + \sum g_i(c)^2 = 0$, contradicting that K is formally real. QED

Now we want to prove an interesting characterization of a real closed field. One should note that the characteristic does not come in the statement.

THEOREM 6.1.10 A field K is real closed if and only if $\sqrt{-1} \notin K$ and $K(\sqrt{-1})$ is algebraically closed.

PROOF: The only if part is contained in Theorem 6.1.9. As for the if part, the conditions imply that irreducible polynomials in X over K have degrees at most 2. We want to show that $a,b \in K$ implies $\sqrt{a^2 + b^2} \in K$. We may assume that $ab \neq 0$. $g(X) = (X^2 - a)^2 + b^2$ is reducible because $\deg g = 4$. The squares of roots of $g(X)$ are $a \pm b\sqrt{-1}$. Let c_1, c_2, d_1, d_2 be the roots of $g(X)$; we may assume that $c_i^2 = a + b\sqrt{-1}$, $d_i^2 = a - b\sqrt{-1}$ $(i = 1, 2)$. Since $c_i^2, d_i^2 \notin K$, we have $c_i, d_i \notin K$. $c_1c_2 = -c_1^2 \notin K$, and therefore c_2 is not a conjugate of c_1. Thus we may assume that d_1 is a conjugate of c_i. Then $c_1^2d_1^2 = a^2 + b^2$ implies that $c_1d_1 = \pm\sqrt{a^2 + b^2} \in K$. If K is not formally real, then $-1 = a_1^2 + \cdots + a_m^2$ $(a_i \in K)$. This and the fact just proved imply that $\sqrt{-1} \in K$, contradicting our assumption. Therefore K is real closed by Theorem 6.1.9. QED

COROLLARY 6.1.11 If K is a subfield of a real closed field L, then the algebraic closure K^* of K in L is real closed.

PROOF: It suffices to show that $K^*(\sqrt{-1})$ is algebraically closed. Since $L(\sqrt{-1})$ is algebraically closed, every element which is algebraic over K is written as $a + b\sqrt{-1}$ $(a,b \in L)$. K is fixed under the automorphism σ of $L(\sqrt{-1})$ over L such that $\sqrt{-1}^{\sigma} = -\sqrt{-1}$, and therefore $a - b\sqrt{-1}$ is also algebraic over K. Therefore

$(a + b\sqrt{-1}) \pm (a - b\sqrt{-1})$ are algebraic over K and $a,b \in K^*$. Thus $K^*(\sqrt{-1})$ is algebraically closed. QED

In order to prove a generalization of Theorem 6.1.10, we prove the following:

LEMMA 6.1.12 If a field K of characteristic $p \neq 0$ has a cyclic extension K_1 of degree p, then for every natural number n, there is a sequence $K = K_0 \subset K_1 \subset \cdots \subset K_n$ of fields K_i such that each K_i is a cyclic extension of degree p of K_{i-1} $(i = 1, 2, \ldots, n)$.†

PROOF: Let f be the mapping such that $fx = x^p - x$. We note that (*) $X^p - X - b$ $(b \in K)$ is irredubible over K if and only if $b \notin fK$ (Theorem 3.8.10). We want to prove that if $a \notin K$, $fa = b \in K$, then $ba^{p-1} \notin f(K(a))$. Assume that $ba^{p-1} \in f(K(a))$. This means that $fd = ba^{p-1}$ by some $d = c_0 + c_1a + \cdots + c_{p-1}a^{p-1}$ $(c_i \in K)$. Noting that $a^p = a + b$, we have $ba^{p-1} = d^p - d = \sum c_i{}^p a^{ip} - \sum c_i a^i = \sum c_i{}^p(a + b)^i - \sum c_i a^i$. Since $1, 2, \ldots, a^{p-1}$ are linearly independent over K, we have $b = c_{p-1}^p - c_{p-1}$ by comparing the coefficients of a^{p-1}. This contradicts the fact that $a \notin K$ [in view of (*) above, $b \in fK$ implies $a \in K$]. Now, starting with a_1 such that $f(a_1) = b_1 \in K$, $a_1 \notin K$, we construct a sequence of a_i and b_i such that $f(a_{i+1}) = b_{i+1} = b_i a_i{}^{p-1}$. Then $K_{i+1} = K_i(a_{i+1})$ gives us the required sequence. QED

Now we state a generalization of Theorem 6.1.10.

THEOREM 6.1.13 Let K^* be the algebraic closure of a field K. If $1 < [K^* : K] < \infty$, then K is real closed, and $K^* = K(\sqrt{-1})$.

PROOF: If $K^* = K(\sqrt{-1})$, then we see the assertion by Theorem 6.1.10, and we assume that $K(\sqrt{-1}) \neq K^*$. Assume first that K is of characteristic p and not perfect. Then $K^{p^{-1}} \cong K$ (Exercise 3.3.5) and therefore $K \subset K^{p^{-1}} \subset K^{p^{-2}} \subset \cdots$ and $[K^* : K] = \infty$,

† It is known that there is a cyclic extension of degree p^n is this case.

which is not the case. Thus K is perfect and, K^* is a Galois extension of $K(\sqrt{-1})$. Let G be the Galois group. Let H be a cyclic subgroup of G of a prime order q. Let L be the subfield corresponding to H. Then K^* is a cyclic extension of L of degree q. If $q =$ (the characteristic), then Lemma 6.1.12 gives us a contradiction. Thus $q \neq$ (the characteristic). Let ζ_n be a primitive q^n-th root of unity. Since $[L(\zeta_1) : L] < q$, we have $\zeta_1 \in L$. Therefore $K^* = L(a^{1/q})$ with $a \in L$ (Theorem 3.8.8). Let b be a root of $X^{q^2} - a$. Then $X^{q^2} - a = \Pi_{i=1}^{q^2} (X - \zeta_2{}^i b)$. Since $[K^* : L] = q$, this polynomial is reducible over L. Let $h(X)$ be a monic irreducible factor which has b as a root. Then since $b \notin L$, $\deg h(X) = q$. Then the constant term c of $h(X)$ is of the form $b^q \zeta_2{}^s$. Since b^q is a root of $X^q - a$, we have $b^q \notin L$. Since $c \in L$, we have $\zeta_2{}^s \notin L$. Thus $K^* = L(\zeta_2)$. Let P be the prime field, and take a natural number r such that $\zeta_r \in P(\zeta_2)$, $\zeta_{r+1} \notin P(\zeta_2)$; such r exists because $[P(\zeta_r) : P]$ is very large if so is r. Let $g(X)$ be the minimal polynomial for ζ_{r+1} over L. $g(X)$ is a factor of $\Pi(X - \zeta_{r+1}{}^i)$, and therefore $g(X) \in P(\zeta_{r+1})[X]$ and $g(X) \in (L \cap P(\zeta_{r+1}))[X]$. Since $\deg g(X)= q$, we have $[P(\zeta_{r+1}) : (L \cap P(\zeta_{r+1})]$ $= q$. On the other hand, since $\zeta_{r+1} \notin P(\zeta_r) = P(\zeta_2)$, $[P(\zeta_{r+1}) : P(\zeta_r)] = q$. Since $r \geq 2$, $\zeta_r \notin L$ and $P(\zeta_r) \not\subseteq L$. Thus $P(\zeta_r) \neq L \cap P(\zeta_{r+1})$. This shows that the commutative group $G^* = G(P(\zeta_{r+1})/P)$ (Theorem 3.8.2) has at least two cyclic subgroups of order q, which implies that G^* is not cyclic and P is not a finite field (Theorem 3.8.3). Thus the characteristic of K is zero, and P is the rational number field. Furthermore, G^* is isomorphic to the unit group of $Z/q^{r+1}Z$ (Theorem 3.8.5), and therefore if q is odd then G^* is cyclic (Exercise 3.8.2). Therefore $q = 2$. Then $\zeta_3 \notin P(\zeta_2)$ and $r = 2$; $\zeta_2 = \pm\sqrt{-1} \in L$, contradicting $\zeta_r \notin L$. Thus K^* must be $K(\sqrt{-1})$. QED

6.2 REAL CLOSURES

Although we saw in the last section that for a given ordered field K, there is a real closed field K' which is algebraic over K, we do not define K' as a real closure of K. We define a real closure of an ordered field K as a field K^* which is real closed, algebraic over K, and the unique order of K^* (cf. Theorem 6.1.8) is an extension of the given order on K. Before proving the existence and uniqueness of real closures, we prove:

LEMMA 6.2.1 Let K be an ordered field, and let K' be the field generated by square roots of all positive elements of K. Then K' is formally real.

PROOF: If there is a nontrivial relation $\sum a_i^2 = 0$ $(a_i \in K')$, then there are positive elements $b_1, \ldots, b_m$ such that all the a_i are in $K'' = K(\sqrt{b_1}, \ldots, \sqrt{b_m})$. Then it suffices to prove that $\sum_{i=1}^n c_i d_i^2 = 0$, $0 < c_i \in K$, $d_i \in K''$ imply $d_i = 0$ for all i. We want to prove this by induction on m. Write $d_i = g_i + h_i\sqrt{b_m}$ $(g_i, h_i \in K''' = K(\sqrt{b_1}, \ldots, \sqrt{b_{m-1}}))$. Then $\sum c_i g_i^2 + b_m(\sum h_i^2) + 2(\sum g_i h_i)\sqrt{b_m} = 0$. Then $\sum c_i g_i^2 + \sum b_m h_i^2 = 0$, and the induction hypothesis implies that $g_i = h_i = 0$ for all i. Thus $d_i = 0$. QED

THEOREM 6.2.2 If K is an ordered field, then there is a real closure K^* of K, and K^* is unique within K-isomorphisms [i.e., if K'' is another real closure of K, then there is an K-isomorphism ψ of K^* to K''. Note here that ψ is an order-isomorphism (cf. Theorem 6.1.8)].

PROOF: Let K' be as in Lemma 6.2.1. Since K' is formally real, there is a real closed field K^* containing K' and algebraic over K' (Theorem 6.1.4). Positive elements of K are squares of elements of K^*. Let P and P^* be the set of positive elements in K and K^*. Our construction shows that $P = P^* \cap K$. Therefore the unique order on K^* is an extension of the given order on K. Thus K^* is a real closure of K. The uniqueness of K^* follows from:

THEOREM 6.2.3 Let K, L be ordered fields, and let K^*, L^* be their real closures. If ψ is an order-isomorphism of K to L, then ψ is extended to an isomorphism of K^* to L^*. Such an extension is unique. In particular, any K-automorphism of K^* is the identity mapping.

For the proof, we need some preparations.

Lemma 6.2.4 Let a, b be elements of a real closed field K (with a unique order). Let $f(X) \in K[X]$. If $f(a) > 0$, $f(b) < 0$, there is $c \in K$ such that $f(c) = 0$; $a > c > b$ or $a < c < b$.

PROOF: We may assume that f is monic. $f(X) = \Pi_{i=1}^{r}(X - c_i) \times \Pi_{j=1}^{s} g_j(X)$ with $c_i \in K$, $g_j(X) = (X + d_j)^2 + e_j$ $(d_j, e_j \in K$, $e_j > 0)$ [Theorem 6.1.10 shows that an irreducible polynomial over K is of degree at most 2; if $e_j < 0$, then $e_j = -e_j^{*2}$ and $g_j(X)$ is reducible]. Then $g_j(a)$, $g_j(b)$ are all positive. We may assume that $c_1 \geq c_2 \geq \cdots \geq c_r$. Let t, u be such that $c_t > a > c_{t+1}$, $c_u > b > c_{u+1}$. Then t is even and u is odd. Thus there is c_i which is in between a and b. QED

In general, for a polynomial $f(X) = \sum_{i=1}^{n} c_i X^i$ over a field K, the <u>standard sequence</u> is defined as follows: It starts with $f_0 = f(X)$ and its derivative $f_1(X) = \sum ic_i X^{i-1}$. When $f_0, \ldots, f_i$ are defined, let r_i be such that $f_{i-1} = f_i q_i - r_i$ $(q_i, r_i \in K[X]$, $\deg r_i < \deg f_i)$. If $r_i = 0$, then f_i is the last term of the standard sequence; otherwise, $f_{i+1} = r_i$.

A sequence of polynomials $g_0(X), \ldots, g_m(X)$ over an ordered field K is called a <u>Sturm sequence</u> on an interval $[a,b]$ $(a,b \in K$, $a < b)$ if the following four conditions are satisfied: (1) there is no c such that $g_m(c) = 0$, $a \leq c \leq b$; (2) $g_0(a)g_0(b) \neq 0$; (3) if $g_i(c) = 0$, $a \leq c \leq b$, $0 < i < m$, then $g_{i-1}(c)g_{i+1}(c) < 0$; and (4) if $g_0(c) = 0$, $a < c < b$, then there are $d_1, d_2 \in K$ such that (i) $d_1 < c < d_2$ and (ii) $g_0(x)g_1(x) < 0$, $g_0(y)g_1(y) > 0$ for any $x,y \in K$ such that $d_1 \leq x < c < y \leq d_2$.

Let $a_0, \ldots, a_m$ be a given sequence of elements of an ordered field K. Let $a_0', \ldots, a_n'$ be a new sequence obtained from $a_0, \ldots, a_m$ by deleting 0 among them (not changing the order). Then count the number of i such that $a_{i-1}'a_i' < 0$ $(1 \leq i \leq n)$. The number is called the number of <u>sign variations</u> of the sequence $a_0, \ldots, a_m$.

Now we state the <u>Sturm theorem</u>:

THEOREM 6.2.5 Let $f(X)$ be a polynomial in X over a real closed field K, and let $f_0 = f, f_1, \ldots, f_s$ be the standard sequence for $f(X)$. For each $c \in K$, let $V_c = V_c(f)$ be the number of sign variations of the sequence $f_0(c), \ldots, f_s(c)$. If $a < b$, $f(a)f(b) \neq 0$, then $\#\{c \in K \mid f(c) = 0,\ a < c < b\} = V_a - V_b$.

PROOF: By the Euclid algorithm, f_s is the largest common factor of f, f_1 and therefore $f_s(a)f_s(b) \neq 0$; furthermore if c is an m-ple root of f, then c is an $(m - 1)$-ple root of f_s. Consider the sequence of $g_i = f_i/f_s$ $(i = 0, 1, \ldots, s)$. These properties show that V_a, V_b are unchanged even if we replace f_i with g_i and also that roots of f are exactly the roots of g_0. Thus we consider $g_0, g_1, \ldots, g_s$ $(= 1)$ instead of the standard sequence. We want to show that $g_0, \ldots, g_s$ form a Sturm sequence. Indeed, (1) is obvious because $g_s = 1$. (2) is assumed. As for (3), since $f_{i-1} = f_iq_i - f_{i+1}$, we have $g_{i-1} = g_iq_i - g_{i+1}$. So, if $g_i(c) = 0$, then since g_i and g_{i+1} have no proper common factor (because $g_s = 1$), we see that $g_{i-1}(c) = -g_{i+1}(c) \neq 0$ and we have (3). As for (4), assume that $g_0(c) = 0$. By our construction, g_0 has no multiple root, and $g_0 = (X - c)h(X)$ with $h(X) \in K[X]$ such that $h(c) \neq 0$. If $f(X) = (X - c)^e h'(X)$ $[h'(c) \neq 0]$, then $f_1 = e(X - c)^{e-1}h'(X) + (X - c)^e \cdot \frac{d}{dx} h'(X)$ and $f_s = (X - c)^{e-1}h_s(X)$ $[h_s(c) \neq 0]$. Therefore we have $g_1(c) = eh(c)$, and we also have the fact that $g_1(c) \neq 0$. Since there are only a finite number of roots of $h(X)g_1(X)$, there are $d_1, d_2 \in K$ such that $d_1 < c < d_2$ and such that $h(X)g_1(X)$ has no root in between d_1 and d_2. Then Lemma 6.2.4 implies that each of h, g_1 does not change positivity nor negativity in $\{x \in K \mid d_1 \leq x \leq d_2\}$. Since $eh(c) = g_1(c)$, we see that $h(c)g_1(c) > 0$ and we have (4). Thus $g_0, \ldots, g_s$ form a Sturm sequence. Therefore, our assertion follows from the following:

LEMMA 6.2.6 Let $g_0, g_1, \ldots, g_s$ be a Strum sequence on an interval $[a,b]$, and let V_c be the number of sign variations of $g_0(c), g_1(c), \ldots, g_s(c)$ for each c of the real closed field K of reference. Then $\#\{c \in K \mid g_0(c) = 0,\ a < c < b\} = V_a - V_b$.

PROOF: Consider $a = a_0 < a_1 < \cdots < a_n = b$ such that $g_0(a_i) \neq 0$ for all i. Then $V_a - V_b = \sum_{i=1}^n (V_{a_{i-1}} - V_{a_i})$, and it suffices to prove the assertion for each interval $[a_{i-1}, a_i]$. Thus we may assume furthermore that (i) if c, c' $(a < c \le c' < b)$ are roots of g_0, then $c = c'$; (ii) if d is a root of some g_i, then either d is a root of g_0 or $d = a$ or b; and (iii) if c $(a < c < b)$ is a root of g_0, then a, b satisfy the condition for d_1, d_2 in (4) above. We also employ an induction on s.

(I) Assume that g_0 has a root c $(a < c < b)$. By Condition (4), $g_0(a)g_1(a) < 0$, $g_0(b)g_1(b) > 0$, and the number of sign variations of $g_0(x)$, $g_1(x)$ is 1, 0 for $x = a$, b, respectively. Thus, if $s = 1$, we finish the proof. Assume $s > 1$. By Assumption (iii), $g_1(x) \ne 0$ if $x \ne c$ $(a \le x \le b)$. By Condition (3), $g_1(c) \ne 0$. Therefore g_1 has no root x $(a \le x \le b)$. Then we see that $g_1, \ldots, g_s$ is a Sturm sequence on the interval $[a,b]$; therefore, by our induction on s, the numbers of sign variations of $g_1(a), \ldots, g_s(a)$ and $g_1(b), \ldots, g_s(b)$ are the same. This proves the present case.

(II) Assume that g_0 has no root in the interval $[a,b]$. If some g_j $(1 \le j < s)$ has no root, then we apply our induction to Sturm sequences $g_0, \ldots, g_j$ and $g_j, \ldots, g_s$, and the assertion is proved. Therefore we assume that for $j = 1, 2, \ldots, s - 1$, it holds that $g_j(a)g_j(b) = 0$. If $s = 1$, then obviously $V_a = V_b$, and we assume that $s > 1$. We want to show that $V_a = V_b$. If $g_j(a) = g_j(b) = 0$ for some j, then $g_{j-1}(a)g_{j+1}(a)$ and $g_{j-1}(b)g_{j+1}(b)$ are negative, and therefore both V_a and V_b remain the same even if we omit g_j (since $g_j(x) \ne 0$ for $a < x < b$, $g_0, \ldots, g_{j-1}, g_{j+1}, \ldots, g_s$ is a Sturm sequence on $[a,b]$), and such a case is proved by induction. Thus we assume that for each i $(1 \le i < s)$, one and only one of $g_i(a)$, $g_i(b)$ is zero. Assume that $s = 2$. If $g_1(a) = 0$, then $g_0(x)g_2(x) < 0$ for $x = a$, and hence for any $a \le x \le b$. Therefore we have $V_a = V_b = 1$. The case where $g_1(b) = 0$ is proved similarly. Assume now that $s > 2$. If we divide the interval $[a,b]$ to $[a,c]$, $[c,b]$ $(a < c < b)$, then, for instance, if $g_1(a) = 0$, then (i) $g_2(a) \ne 0$ and therefore g_2 has no root in between a and c and (ii) g_1 has no root in between c and b. Therefore we reduce s on each interval, and we complete the proof. QED

LEMMA 6.2.7 Consider a polynomial $f(X) = X^n + c_1X^{n-1} + \cdots + c_n$ over an ordered field K. We denote by $|x|$ $(x \in K)$ the non-negative element among $\pm x$. Let m be $\max\{1, |c_1| + |c_2| + \cdots + |c_n|\}$. If an element a of K is a root of $f(X)$, then $-m \leq a \leq m$.

PROOF: If $|a| \leq 1$, then there is nothing to prove. Assume that $a > 1$. Since $a^n + c_1a^{n-1} + \cdots + c_n = 0$, we have $|-a| = |c_1 + c_2a^{-1} + \cdots + c_na^{-(n-1)}| \leq |c_1| + |c_2a^{-1}| + \cdots + |c_na_n^{-(n-1)}| \leq |c_1| + \cdots + |c_n|$. QED

COROLLARY 6.2.8 Under the notation in Theorem 6.2.5, assume that $f(X)$ has no multiple root. Let $b_1 < b_2 < \cdots < b_s$ be all of mutually distinct roots of $f_1(X)$ $[= \frac{d}{dx} f(X)]$ in K, and let $m \in K$ be such that all roots of $f(X)$ are in between $-m$ and m as was given in lemma 6.2.7. Also let $b_0, b_{s+1} \in K$ be such that $b_0 < \min\{-m, b_1\}$, $b_{s+1} > \max\{m, b_s\}$. Then the number of roots of $f(X)$ in K is at most $s + 1$; as a matter of fact, $f(X)$ has at most one root c such that $b_{i-1} < c < b_i$.

PROOF: Since $f(X)$ has no multiple root, the standard sequence $f_0 = f, f_1, \ldots, f_s$ for f is a Sturm sequence. Since $f(b_i) \neq 0$, there are $d_i, e_i \in K$ such that $d_0 = b_0 < e_1 < b_1 < d_1 < e_2 < b_2 < \cdots < d_{s-1} < e_s < b_s < d_s < b_{s+1} = e_{s+1}$ and such that $f(X)$ has no root in between e_i and d_i $(i = 1, \ldots, s)$. Since $f_1(X)$ has no root in between d_{i-1} and e_i, $V_{d_{i-1}}(f_1) = V_{e_i}(f_1)$ (Lemma 6.2.6) and therefore $V_{d_{i-1}}(f) - V_{e_i}(f) \leq 1$. QED

PROOF of THEOREM 6.2.3: (I) Let $f(X)$ be an arbitrary polynomial in X over K. If $f_0, \ldots, f_s$ is the standard sequence for $f(X)$, then $\psi f_0, \ldots, \psi f_s$ is the standard sequence for ψf. Since $V_a(f) - V_b(f) = V_{\psi a}(\psi f) - V_{\psi b}(\psi f)$, we see that the numbers of mutually distinct roots of $f(X)$ and $\psi f(X)$ in K^* and L^* are the same.

(II) We want to show that if $a_1 < a_2 < \cdots < a_n$ are elements of K^*, then ψ is extended to an injection of $K(a_1, \ldots, a_n)$ into L^* (not as an ordered field) such that $\psi a_1 < \psi a_2 < \cdots < \psi a_n$.

Indeed, $a_i' = \sqrt{a_i - a_{i-1}} \in K^*$ $(i = 2, \ldots, n)$. Set $K_1 = K(a_1, \ldots, a_n, a_2', \ldots, a_n')$. Take a^* such that $K_1 = K(a^*)$ (Theorem 3.4.1), and let $f(X)$ be the minimal polynomial for a^*. Then by (I) above, $\psi f(X)$ has a root, say a'' in L^*. Then we have an isomorphism ϕ of $K(a^*)$ to $L(a'')$ which extends ψ and such that $\phi(a^*) = a''$. Then $L(a'')$ contains $\phi(a_i')$, and $0 < \phi(a_i')^2 = \phi a_i - \phi a_{i-1}$ as we wished.

(III) Now we define a mapping σ of K^* to L^* as follows: For each $a \in K^*$, let $f_a(X)$ be the minimal polynomial for a over K, and take all roots $a_1 < a_2 < \cdots < a_n$ of $f_a(X)$ in K^*. Then, let $b_1 < \cdots < b_n$ be all the roots of $\psi f_a(X)$ in L^* [by (I) above]. Taking j such that $a = a_j$, we define σa to be b_j. (i) If $a \in K$, then $f_a(X) = X - a$ and $\sigma a = a$. (ii) (I) above shows that σ gives a one-one correspondence between K^* and L^*. (iii) Let $a,b \in K^*$, and consider minimal polynomials $f_a(X)$, $f_b(X)$, $f_{a+b}(X)$, and $f_{ab}(X)$. Let $c_1 < c_2 < \cdots < c_m$ be all mutually distinct roots of $f_a(X)f_b(X)f_{a+b}(X)f_{ab}(X)$. By (II) above, ψ is extended to an injection ϕ of the field $K(c_1, \ldots, c_m)$ into L^* so that $\phi c_1 < \phi c_2 < \cdots < \phi c_m$. Let d be any one of $a + b$ and ab. If the roots of $f_d(X)$ are $c_{i_1} < c_{i_2} < \cdots < c_{i_n}$ and if $d = c_{i_j}$, then the roots of $\psi f_d(X)$ are $\phi c_{i_1} < \cdots < \phi c_{i_n}$ (because of the property of ϕ). Therefore our definition of σ implies that $\sigma d = \phi d$. Since ϕ is an injection of a field, we see that σ is an isomorphism of a field. (iv) If $0 < a \in K^*$, then $a = b^2$ with $b \in K^*$ and $\sigma a = \sigma b^2$. Thus $a > 0$ implies $\sigma a > 0$, and σ is an order-isomorphism. (v) If we want to extend ψ to an injection τ of K^* into L^*, then τ must satisfy the condition in the definition of σ, and therefore the extension is unique. QED

We note here that it follows immediately from Corollary 6.1.11 that:

LEMMA 6.2.9 If an ordered field K is a subfield of a real closed field L, not just as a field but as an ordered field, then the algebraic closure K^* of K in L is the real closure of K.

THEOREM 6.2.10 Let K be an ordered field, K' an algebraic extension of K, and L a real closed field containing K as an ordered field. Let H be the set of K-injections of the field K' into L. (1) If $\sigma \in H$, then by defining an order by $a \geq b$ if and only if $\sigma a \geq \sigma b$, K' becomes an ordered field. Conversely, (2) if the order on K is extended to K' so that K' is an ordered field, then the order is obtained as in (1) by one and only one σ in H.

PROOF: (1) is immediate. As for (2), let K^* be the real closure of K'. Then K^* is the real closure of K, and there is an K-isomorphism ψ of K^* to the algebraic closure K'' of K in L. Its restriction σ on K' gives an element of H. If $\sigma, \tau \in H$ give the same order on K', then $\sigma a \rightsquigarrow \tau a$ $(a \in K')$ gives an isomorphism ϕ of $\sigma K'$ to $\tau K'$. ϕ is extended to an isomorphism ϕ' of K'' to itself. ϕ' must be identity (Theorem 6.2.3), and $\sigma = \tau$. QED

6.3 THE 17-TH PROBLEM OF HILBERT

The following problem is called the <u>17-th problem of Hilbert</u>†:

Let $f(X_1,\ldots,X_n)$ be a rational function in $X_1, \ldots, X_n$ with rational coefficients, i.e., $f = g/h$ with $g,h \in Q[X_1,\ldots,X_n]$, $h \neq 0$. Assume that for any $a_1, \ldots, a_n \in Q$ such that $h(a_1,\ldots,a_n) \neq 0$, it holds that $f(a_1,\ldots,a_n) \geq 0$. Does it follow that $f = \sum g_i^2$ with suitable $g_i \in Q(X_1,\ldots,X_n)$?

The problem was answered by E. Artin in a generalized form, and we introduce his proof and related results.

THEOREM 6.3.1 Let K be a field of characteristic different from 2, and let a be an element of K. It is impossible to make K into an ordered field so that $a < 0$ if and only if there are $b_1, \ldots, b_n$ in K such that $a = \sum b_i^2$.

PROOF: The if part is obvious. Assume the nonexistence of such

---

† D. Hilbert presented 23 problems at the International Congress of Mathematicians at Paris in 1900, and this is the 17-th of these problems.

b_i. Let F be the set of subfields K' of the algebraic closure K^* of K such that $K \subseteq K'$ and a is not a sum of square elements in K'. Then F is an inductive set and has a maximal member L^*. Lemma 6.1.5 implies that L^* is formally real, and therefore L^* is made to an ordered field. We want to show that $a < 0$ by the order. It suffices to show that $-a$ is a square element in L^*. Assume the countrary. Then $L^*(\sqrt{-a}) \neq L^*$, and therefore there are $c_1, \ldots, c_m \in L^*(\sqrt{-a})$ such that $a = \sum c_i^2$. Write $c_i = a_i + b_i\sqrt{-a}$ $(a_i, b_i \in L^*)$. Then $a = \sum a_i^2 - a\sum b_i^2$ and $a(1 + \sum b_i^2) = \sum a_i^2$. Then Lemma 6.1.5 implies that a is a sum of square elements in L^*, which contradicts $L^* \in F$. QED

In the following, throughout this section, we fix the following notation: K is an ordered field with real closure K^*. $x_1, \ldots, x_n$ are algebraically independent elements over K, and we set $L = K(x_1, \ldots, x_n)$. L becomes an ordered field (by Lemma 6.3.2 below), and L^* is the real closure of L. $(x_1, \ldots, x_n)$ may be written as (x); if $a_1, \ldots, a_n$ are elements of K, then $(a_1, \ldots, a_n)$ may be written as (a). Elements of L may be written as $g(x)$, $h(x)$, etc., and polynomials in Y over L may be written as $f(x;Y)$, $g(x;Y)$, etc.. Then, for instance, $f(a;Y)$ is the polynomial in Y which is obtained from $f(x;Y)$ by setting $x_i = a_i$.

LEMMA 6.3.2 The order of K can be extended to $L = K(x_1, \ldots, x_n)$ so that L is an ordered field.

PROOF: $K^*(x_1, \ldots, x_n)$ is formally real (Exercise 6.1.5) and is made into an ordered field; it is an extension of K^* as an ordered field by Theorem 6.1.8. The restriction of the order to L is the required. QED

Let $g_i(x) \in L$ $(i = 1, \ldots, m)$. We say that $(g_1(a), \ldots, g_m(a))$ <u>has the same signs</u> as $(g_1(x), \ldots, g_m(x))$ if $a_j \in K$ are such that all $g_i(a)$ is well defined (i.e., denominators do not vanish) and if $g_i(a) > 0$, $= 0$, < 0 according as $g_i(x) > 0$, $= 0$, < 0 under the given order of L $(i = 1, \ldots, m)$.

LEMMA 6.3.3 If $f(x;Y)$ is a monic polynomial in Y over L

and if r is the number of mutually distinct roots of $f(x;Y)$ in L^*, then there are a natural number m and elements $g_i(x)$ of L $(i = 1, \dots, m)$ such that it holds that if $a_j \in K$ and if $(g_1(a), \dots, g_m(a))$ has the same signs as $(g_1(x), \dots, g_m(x))$, then r is the number of mutually distinct roots of $f(a;Y)$ in K^*.

PROOF: Write $f = Y^s + c_1Y^{s-1} + \cdots + c_s$ $[c_i = c_i(x) \in L]$, and set $h = 1 + \sum |c_i|$ $(|c_i| = \pm c_i,\ |c_i| > 0)$. Let $f_0 = f$, f_1, $\dots$, f_t be the standard sequence for f as a polynomial in Y. Let $g_1, \dots, g_m$ be all of the following: (i) All coefficients of f_0, $f_1, \dots, f_t$ (as polynomials in Y), (ii) all $f_i(x;H)$ with $H = \pm h$, and (iii) all $f_i(x;H)/f_j(x;H)$ with i, j such that $j > i$, $H = \pm h$, and $f_i(x;H) \neq 0$. Lemma 6.2.7 and Theorem 6.2.5 show that $V_{-h}(f) - V_h(f) = r$. By our choice of the g_i, $V_{-h(a)}(f(a;Y)) = V_{-h}(f)$, $V_{h(a)}(f(a;Y)) = V_h(f)$. Since c_j are among g_i, $-h(a)$ and $h(a)$ bound the roots of $f(a;Y)$ in K^*, and we see the result. QED

LEMMA 6.3.4 Assume that $f_1(x;Y), \dots, f_s(x;Y)$ are monic polynomials in Y over L such that $b_1 < b_2 < \cdots < b_s$ with suitable root b_i of $f_i(x;Y)$ in L^* $(i = 1, \dots, s)$. Then there are a natural number m and elements $g_i(x)$ of L $(i = 1, \dots, m)$ such that it holds that if $a_j \in K$ and if $(g_1(a), \dots, g_m(a))$ has the same signs as $(g_1(x), \dots, g_m(x))$, then $b_1^* < \cdots < b_s^*$ with suitable root b_i^* of $f_i(a;Y)$ in K^* $(i = 1, \dots, s)$.

PROOF: Let M be the field generated by b_i and $\sqrt{b_{j+1} - b_j}$ over L $(i = 1, \dots, s;\ j = 2, \dots, s)$. Then $M = L(y)$ with an element y (Theorem 3.4.1). Let $g(x,Y) = Y^t + c_1Y^{t-1} + \cdots + c_t$ $[c_i = c_i(x) \in L]$ be the minimal polynomial for y over L. Take $b_i'(x;Y)$, $e_j'(x;Y)$, $e_j''(x;Y) \in L[Y]$ such that $b_i = b_i'(x;y)$, $\sqrt{b_{j+1} - b_j} = e_j'(x;y)$, $e_j'(x;y)e_j''(x;y) = 1$. $F_i(x;Y) = f_i(x;b_i'(x;Y))$ is divisible by $g(x;Y)$, and $F_i = g(x;Y)f_i'(x;Y)$ with $f_i' \in L[Y]$. Since $b_{j+1} - b_j = e_j'(x;y)^2$, we have $b_{j+1}'(x;Y) - b_j'(x;Y) - e_j'(x;Y)^2 = g(x;Y)h_j(x;Y)$ with $h_j \in L[Y]$. Similarly, $e_j'(x;Y)e_j''(x;Y) - 1 = g(x;Y)h_j'(x;Y)$. Let $g_1, \dots, g_m$ be all of (i) the coefficients of $f_i(x;Y)$, $f_i'(x;Y)$, $g(x;Y)$, $h_j(x;Y)$, $h_j'(x;Y)$, $b_i'(x;Y)$, $e_j'(x;Y)$, $e_j''(x;Y)$ (as polynomials in Y) and (ii) the finite number of polynomials obtained for $g(x;Y)$ by applying Lemma 6.3.3. Then, for $(a) = (a_1,\ldots,a_n)$ satisfying the assumption in the statement of our assertion, since $g(x;Y)$ has a root y in L^*, Lemma 6.3.3 shows that $g(a;Y)$ has a root, say a^*, in K^*. Then $f_i(a;b_i'(a,a^*)) = f_i'(a;a^*)g(a;a^*) = 0$. This shows that $b_i^* = b_i'(a;a^*)$ is a root of $f_i(a;Y)$. $e_j'(a;a^*)e_j''(a;a^*) = 1$ and $e_j'(a;a^*) \neq 0$. Thus $b_{j+1}^* - b_j^* = e_j'(a;a^*)^2 > 0$. QED

THEOREM 6.3.5 Assume furthermore that K is order-isomorphic to a subfield of the real number field R. If $d(x)$, $f_1(x)$, ..., $f_r(x) \in L$ are given, and if $d(x) \neq 0$, then there are rational numbers a_1, ..., a_n such that $(f_1(a),\ldots,f_r(a))$ and $(f_1(x),\ldots,f_r(x))$ have the same signs and $d(a) \neq 0$.

PROOF: We employ an induction on n. If $n = 0$, then the assertion is obvious. We assume that the assertion is true for n and consider the case of $n + 1$. (x) and (a) will mean $(x_1,\ldots,x_n)$, $(a_1,\ldots,a_n)$, respectively; x_{n+1} is denoted by Y. Thus $d(x)$, $f_i(x)$ above is denoted by $d(x,Y)$, $f_i(x;Y)$. If some f_i is not in $L[Y]$, then multiplying the square of the denominator to f_i and also to $d(x;Y)$, we may assume that $f_i \in L[Y]$. Then, we may replace these f_i with all monic irreducible factors of these f_i and coefficients of the highest degree terms of f_i. Thus we may assume that each f_i is either an element of L or an irreducible monic polynomial in Y. Let $b_1 < b_2 < \cdots < b_s$ be all roots of $\Pi\, f_i(x;Y)$ in L^*. Then renumbering f_i (and repeating the same if necessary), we may assume that (i) if $i \leq s$, then b_i is a root of $f_i(x;Y)$ and (ii) if $i > s$, then $f_i \in L$. Let $g_1(x)$, ..., $g_t(x) \in L$ be all of (1) those $g_j(x)$ obtained for each $f_i(x;Y)$ by applying Lemma 6.3.3, (2) those $g_j(x)$ obtained for f_1, ..., f_s; $b_1 < \cdots < b_s$ by applying Lemma 6.3.4, (3) f_i with $i > s$, and (4) discriminants $d_i(x)$ of f_i $(i \leq s)$. Set $d'(x) = (\Pi\, d_i(x))$(one of nonzero coefficients of $d(x;Y)$). By our induction hypothesis, there are rational numbers a_1, ..., a_n such that $(g_1(a),\ldots,g_t(a))$ and $(g_1(x),\ldots,g_t(x))$ have the same signs and $d'(a) \neq 0$. By (3) above, $(f_{s+1}(a),\ldots,f_r(a))$ and $(f_{s+1}(x),\ldots,f_r(x))$ have the same signs. Let us consider f_i with $i \leq s$. Let the roots of f_i in L^* be $b_{i_1} < \cdots < b_{i_{n(i)}}$. This means that $f_j = f_i$ if and only if $j = i_k$. Apply Lemma 6.3.4, and we have roots $b_1^* < \cdots < b_s^*$ of $f_1(a;Y), \ldots, f_s(a;Y)$. Then $b_{i_1}^* < \cdots < b_{i_{n(i)}}^*$ are roots of $f_i(a;Y)$; they are all roots of $f_i(a;Y)$ in K^* by Lemma 6.3.3. Now, $f_i(x;Y) = \Pi_j (Y - b_{i_j}) \cdot \Pi_k q_{ik}(x;Y)$ with q_{ik} ($\in L^*[Y]$) monic, irreducible, and of degree > 1. Then q_{ik} is of the form $(Y + p)^2 + q$ ($p,q \in L^*$, $q > 0$). By (4) above, $f_i(a;Y)$ has no multiple root, and therefore we have $f_i(a;Y) = \Pi_j (Y - b_{i_j}^*) \cdot \Pi_k q_{ik}(a;Y)$ with irreducible $q_{ik}(a;Y)$ (because only $b_{i_j}^*$ are roots in K^*). Therefore $q_{ik}(a;a_{n+1}) > 0$ for any a_{n+1}. Thus, the signs of $f_i(a;a_{n+1})$, $f_i(x;x_{n+1})$ are determined by j, j' such that $b_u^* < a_{n+1}$ if and only if $u < j$; $b_u < x_{n+1}$ if and only if $u < j'$. Therefore it suffices to take a rational number a_{n+1} such that $j = j'$ and $d(a;a_{n+1}) \neq 0$. QED

The following is the answer by E. Artin to the 17-th problem of Hilbert.

THEOREM 6.3.6 Assume that K is a formally real field such that every ordered field structure of K makes K to be order-isomorphic to a subfield of the real number field R. Then in $L = K(x_1,\ldots,x_n)$ as before, we have the following: For an element $f(x)$ of L, there are $g_1(x), \ldots, g_m(x) \in L$ such that $f(x) = \sum g_i(x)^2$ if and only if $f(a) \geq 0$ for every $(a) = (a_1,\ldots,a_n)$ (a_i rational numbers) such that $f(a)$ is well defined (i.e., the denominator does not vanish by setting $x_i = a_i$ ($i = 1, \ldots, n$)).

PROOF: The only if part is obvious. Assume the nonexistence of such $g_i(x)$. Then Theorem 6.3.1 implies that the order of K can be extended to L so that $f(x) < 0$. Then Theorem 6.3.5 shows that there are $a_1, \ldots, a_n$ such that $f(a) < 0$. QED

6.4 VALUATIONS ASSOCIATED TO AN ORDER

THEOREM 6.4.1 Let S be a subfield of an ordered field K. Set $R = \{x \in K \mid -a < x < a$ for some $a \in S\}$ and $M = \{x \in K \mid -a$

$< x < a$ for any positive a in $S\}$. Then R is a valuation ring of K, and M is its maximal ideal. Let ψ be the natural homomorphism of R onto R/M. Then R/M is naturally an ordered field; $\psi a > 0$ if and only if $a > 0$, $a \notin M$.

This R is called the valuation ring of K with respect to S.

PROOF: If $x \in K$, $x \notin R$, then (i) in case $x > 0$, $x > a$ for any $a \in S$ and $x^{-1} < b$ for any positive $b \in S$ and $x^{-1} \in M$; (ii) the case $x < 0$ is similar. Thus R is a valuation ring of K, and M is its maximal ideal as is easily seen. Set $P' = \{\psi x \mid x > 0, x \notin M\}$. $0 \notin P'$ obviously. $a',b' \in P'$ implies that $a' + b'$, $a'b' \in P'$ (note that if $a > 0$, $b > 0$, $ab \notin M$, then $a > c$, $b > c$ for some positive c in S and $a + b > c$). Obviously if $a' \in R/M$, then it holds one of $a' = 0$, $a' \in P'$, $-a' \in P'$. Thus the order of R/M is well defined (Lemma 6.1.7). QED

Our main aim in this section is to discuss a converse of this. Although we can adapt the following main result to the general S, for the simplicity of the statement, we restrict ourselves to the case where S is the rational number field $\mathbb{Q}$ (cf. Exercise 6.4.2).

THEOREM 6.4.2 Let R be valuation ring of a field K. Assume that a homomorphism ψ of R into the real number field $\mathbb{R}$ is given so that the kernel of ψ is the maximal ideal M. Then one can make K into an ordered field so that (1) R is the valuation ring of the ordered field K with respect to $\mathbb{Q}$, and (2) ψ induces an order-isomorphism of R/M to ψR. Let G be the value group of an additive valuation v defined by R, and set $H = \{2g \mid g \in G\}$. Then such an order for K is unique if and only if $G = H$; in general, $\#(G/H)$ signifies how many orders for K exist.

In order to prove this theorem, we need some preliminaries on maximally complete valuation rings. We say that a valuation ring V of a field L is maximally complete if there is no proper extension L' of L such that some extension of v to L' has the same value group and the same residue class field as v. It is known that a complete discrete valuation ring is maximally complete (this follows from a general theory of complete local rings). However, nondiscrete complete valuation rings need not be maximally complete. As for valuation rings of Krull dimension at least 2, we do not define completeness. The notion of maximal completeness, however, is also applied to such a general case.

Let V be a valuation ring with maximal ideal M. A valuation ring V^* containing V is called a maximal completion of V if (i) every ideal of V^* is generated by elements of V, and (ii) V^*/MV^* is naturally identified with V/M [namely, (i') the value groups coincide and (ii) residue class fields coincide], and (iii) V^* is maximally complete.

LEMMA 6.4.3 Let R be the valuation ring of an additive valuation v of a field K. Consider the residue class field $K' = R/M$ of R and the value group G of v. Then $\#(K) \leq \#(K'^G)$, where K'^G denotes the set of mappings of G into K'.

PROOF: We fix a mapping w of G into K such that $v(wg) = g$ for every $g \in G$. Let K^* be a set of representatives for $K' - \{0\}$. Let G^* be the set of well-ordered subsets of G consisting only of nonnegative elements. We define here a section $S<s>$ of a well-ordered set S by an element s $(\in S)$ to be $\{x \in S \mid x < s\}$. We consider the set F of mappings f of G to K' such that the supports $S = \{x \in G \mid fx \neq 0\}$ are members of G^*. Our convention is that f_S, g_S will denote members of F whose supports are S; when we have fixed an f_S and if S' is a section of S, then $f_{S'}$ denotes the restriction of f_S to S', i.e., $f_{S'}(x) = f_S(x)$ if $x \in S'$, and $f_{S'}(x) = 0$ otherwise. (1) If S is empty, then f_S is the zero-mapping. We associate 0 $(\in R)$ to f_S in this case. (2) The general case is decided inductively. One requirement is that if an element a of R is associated to $f_S \in F$, then to each $f_{S'}$, S' being a section $S<s>$ of S, there must be associated some element b already and $v(a - b) \geq s$. Now consider f_S with nonempty S. If there is a section S' of S such that we decided already that no element of R is associated to $f_{S'}$, then we decide not to associate any element of R to f_S. So we consider the other case.

(i) Assume that S has the largest member m. Then let b be the element associated to $f_{S<m>}$, and let k be the representative of $f_S(m)$. Then we decide to associate $b + k(wm)$ to f_S. (ii) Assume that S has no largest member. Let b_s be the element associated to $f_{S<s>}$ for each $s \in S$. Consider the set of elements a of R such that $v(a - b_s) \geq s$ for every $s \in S$. If this is not empty, we take one a and decide to associate it to f_S; if this is empty, we decide not to associate any element to f_S. Thus association of elements of R to members of F is decided. If suffices to show that every element c of R is associated to some member of F, because $F \subseteq K'^G$.

We note here that if b_S, b_T are associated to f_S, g_T $(\in F)$ $(S,T \in G^*)$ and if $v(b_S - b_T) = r$, then $S' = \{s \in S \mid s < r\}$ coincides with $T' = \{t \in T \mid t < r\}$ and furthermore $f_{S'} = g_{S'}$. Indeed, let U be the set of $u \in S \cap T$ such that $f_S(x) = g_T(x)$ for all $x \leq u$. It suffices to show that $U \supseteq S' = T'$. Assume the contrary. Let m be the least element in $(S' - (U \cap S')) \cup (T' - (U \cap T'))$. We may assume that $m \in S'$. Then $S<m> \subseteq U$ and $S<m> = T<m'>$ with $m' \in T$. Let b_U be the element associated to $f_{S<m>}$. With $S'' = S<m> \cup \{m\}$, we associated $b_U + k(wm)$ to $f_{S''}$ with the representative k of $f_S(m)$. With $T'' = T<m'> \cup \{m'\}$, we associated $b_U + k'(wm')$ to $g_{T''}$ with the representative k' of $g_T(m')$. Therefore our assumption that $v(b_S - b_T) = r > m$ implies that $v(k(wm) - k'(wm')) > m$, and $k = k'$, $m = m'$. This contradicts the choice of m.

Now, let c be an arbitrary element of R, and let F_c be the set of $f_S \in F$ such that some element b_S is associated to f_S and $v(c - b_S) > s$ for every $s \in S$. If $f_S, g_T \in F_c$, then the observation made above shows that either $f_S = g_T$ or one of S, T, say T, is a section of S and $f_T = g_T$. Therefore the union U of all S such that some $f_S \in F_c$ forms a well-ordered set and these f_S define $f_U \in F$ naturally (i.e., each f_S is the restriction of f_U to S). Then c satisfies our condition for an element associated to f_U, and $f_U \in F_c$. Let b_U be the element associated to f_U.

If $b_U \neq c$, then set $g = v(c - b_U)$, $k = (c - b_U)/wg$, and let k' $\in K^*$ be the representative of (k modulo M). Then, with $S = U \cup \{g\}$, define f_S by $f_S(g) = k'$, $f_S(u) = f_U(u)$ for $u \in U$. Then $f_S \in F_c$, which contradicts our choice of U. Thus $b_U = c$. QED

THEOREM 6.4.4 For a given valuation ring R of a field K, there is a maximal completion R^* of R.

PROOF: With the notation as in Lemma 6.4.3, consider $\#(K'^G)$, and let L be an algebraically closed field containing K such that $\#(L) > \#(K'^G)$. In the present proof, we understand by valuation rings, those contained in L only. Let F be the set of valuation rings R' with maximal ideal M' such that the value group and the residue class field of R' coincide with those of R. Then, as is easily seen, F is an inductive set and has a maximal member, say R^*. Then $\#(R^*) < \#(L)$ by Lemma 6.4.3, and therefore L has infinite transcendence degree over R^* (cf. the proof of Theorem 3.11.6). Therefore, if R^* is not maximally complete, then R^* has a proper extension, sharing residue class field and value group, within L. QED

LEMMA 6.4.5 A maximally complete valuation ring R is a Hensel ring.

PROOF: If R is not a Hensel ring, then Theorem 5.11.9 shows that there is a proper unramified extension of R having the same residue class field. QED

LEMMA 6.4.6 Let R with maximal ideal M be a valuation ring of a field K. If R/M is formally real, then so is K.

PROOF: If $X_1^2 + \cdots + X_n^2 = 0$ has a nontrivial solution $(a_1, \ldots, a_n)$ in K, dividing a_j by one a_i with least value, we may assume that one a_i is 1 and all are in R. Then we have a nontrivial solution of the equation in R/M. QED

THEOREM 6.4.7 Let R be the valuation ring of an additive valuation v of a field K. Let M be the maximal ideal of R, and let G be the value group of v. If R is maximally complete, R/M is real closed, and if G is divisible (namely, for any $b \in G$ and any natural number n, there is $c \in G$ such that $nc = b$), then K is real closed.

PROOF: K is formally real by Lemma 6.4.6. Let K'' be a proper algebraic extension of K, and let R'' be the integral closure of R in K''. Lemma 6.4.5 and Theorem 5.9.3 show that R'' is a valuation ring giving an extension of R. Since G is divisible, the ramification exponent must be 1. Since R/M is real closed, the fact that R is maximally complete implies that $R''/MR'' = (R/M)(\sqrt{-1})$. Thus $X^2 + 1$ has a root in R''/MR''. Since R'' is a Hensel ring by Exercise 5.11.2, we have $\sqrt{-1} \in R'' \subseteq K''$. Therefore a real closed field given by Theorem 6.1.4 cannot be a proper extension of K, and K is real closed. QED

PROOF of THEOREM 6.4.2: (1) Existence of an order: For our purpose, we may prove the same for a suitable extension field. Therefore, (i) we may assume that the residue class field is real closed, because, if $f(X)$ is a monic polynomial which is irreducible modulo M, then adjoining a root of $f(X)$ gives an extension of R with degree equal to $\deg f(X)$; (ii) we may assume that the value group G is divisible, because if there is a prime number p and an element $g \in G$ such that $px = g$ has no solution in G, then adjoining $b^{1/p}$, with b such that $vb = g$, gives a solution of $px = g$ in the extended field, and in this case, the ramification exponent is p and therefore the residue class field does not chang (Theorem 5.9.6); and (iii) we may assume furthermore that R is maximally complete, considering a maximal completion of R instead of R. Under these assumptions, K is real closed (Theorem 6.4.7), and therefore K has a unique ordered field structure. Let P be the set of positive elements: $P = \{x^2 \mid x \in K\}$ (Theorem 6.1.6). (i*) Assume that $a \in R$, $\psi a > 0$. If $a < 0$, then $a = -x^2$. $va = 0$ implies $vx = 0$ and $x \in R$, and hence $\psi a = -(\psi x)^2 \leq 0$, a contradiction. Thus $\psi a > 0$ implies $a > 0$. (ii*) Assume that $\psi a = 0$, i.e., $a \in M$. If $a > r$ for some positive rational number r, then $a - r = x^2$ and $-(\psi r) = (\psi x)^2$, a contradiction. Thus $a < r$. Similarly, $-r < a < r$ for any positive rational number r. These (i*), (ii*) show that the valuation ring R' with maximal ideal M' of K with respect to Q contains R and $M \subseteq M'$. Then Theorem 5.7.2 shows that $R' = R_{M' \cap R} = R_M = R$ and $M' = M$.

(2) Choice of orders: We consider the original situation. For each $g \in G$, we choose $b_g \in K$ such that $v(b_g) = g$ and such that if $g \in H$ then b_g is a square element. When K is an ordered field as required, we see that if $a \in R$, $\psi a > 0$, then $a > 0$ by our requirement. By our choice, $b_h > 0$ for $h \in H$. Therefore, if $x \in K$, $vx = h \in H$, then $x > 0$ if and only if $\psi(x/b_h) > 0$. Thus, in case $G = H$, the order of K must be unique. Now consider the case $G \neq H$. Let $\{g'_\lambda \in G/H \mid \lambda \in \Lambda\}$ be a linearly independent base of G/H over the prime field of characteristic 2, and let g_λ be a representative of g'_λ for each λ. Take $a_\lambda \in K$ such that $v(a_\lambda) = g_\lambda$. Then, at Step (ii) in (1), we could adjoin all of $\sqrt{a_\lambda}$. Then we obtain an order of K such that $a_\lambda > 0$. Then the original G is contained in the new H, and the order of K is uniquely determined. But any number of a_λ may be replaced by $-a_\lambda$; and a different choice of such $-a_\lambda$ gives a different order. This proves the last assertion. QED

6.5 FINITELY GENERATED FORMALLY REAL FIELDS

THEOREM 6.5.1 Let K be an ordered subfield of the real number field R such that K cannot be given any other ordered field structure.† Assume that a formally real field L is finitely generated over K. Set $t = \text{trans.deg}_K L$. Let s, s' be nonegative rational integers. Then L is made into an ordered field which extends K such that (i) the valuation ring of L with respect to Q is of Krull dimension s, and (ii) s' is the maximum of the transcendence degree of intermediate ordered fields of K and L which are order-isomorphic to a subfields of R, if and only if either (1) $s = 0$, $s' = t \leq \text{trans.deg}_K R$ or (2) $s \neq 0$, $s' \leq \text{trans.deg}_K R$, $s + s' \leq t$.

PROOF: The only if part is obvious, and we observe the if part. Consider first the real closure L^* of L and the real closure K^*

† Examples of such K are (i) the rational number field, and (ii) a subfield K of R having the property that if $a > 0$, then $\sqrt{a} \in K$.

of K inside L^*. Then, by the uniqueness of K^* (Theorem 6.2.2), K^* can be regarded as a subfield of R. Then, considering $L \vee K^*$ and K^* instead of L and K, we may assume that K is real closed. Let $z_1, \ldots, z_t$ be a transcendence base for L over K, and let b be an integral element over $P = K[z_1,\ldots,z_t]$ such that $L = K(z_1, \ldots,z_t, b)$. Let $f(z;X) = X^m + c_1X^{m-1} + \cdots + c_m$ ($c_i \in P$) be the minimal polynomial for b over P. Then we can apply Theorem 6.3.6 to K and $L' = K(z_1,\ldots,z_t,X)$, and we see that either $f(z;X) = g_1^2 + \cdots + g_v^2$ with $g_i \in L'$ or there are rational numbers $a_1, \ldots, a_t$, d such that $f(a,d) < 0$ [$f(a;X)$ being well-defined in the sense over there]. Since L is formally real and since $f(z;b) = 0$, we cannot have the first case, and we have the second case.

(1) The case where $s = 0$: In this case, $\text{trans.deg}_K R \geq t$, and we can choose real numbers $r_1, \ldots, r_t$ such that (i) they are algebraically independent over K and (ii) $f(r;d) < 0$. Then $f(r;X)$ has a root, say b', in R. Then $(z_1,\ldots,z_t,b) \rightsquigarrow (r_1,\ldots,r_t, b')$ gives an injection of L into R over K.

(2) Assume that $s \neq 0$. Let $D(z)$ be the discriminant of $f(z;X)$. By our proof of Theorem 6.3.6, we may assume that $D(a) \neq 0$. Considering $z_i - a_i$ instead of z_i, we may assume that all a_i are zero. Take real numbers $r_1, \ldots, r_{s'}$ such that they are algebraically independent over K and $f(r_1,\ldots,r_{s'},0,\ldots,0;d) < 0$. Then $f(r_1,\ldots,r_{s'},0,\ldots,0;X)$ has a root b' in R. Now, consider a valuation ring V of $K(z_1,\ldots,z_t)$ such that Krull dim $V = s$, the maximal ideal M contains $z_{s'+1}, \ldots, z_t$, and V/M coincides naturally with $K(z_1,\ldots,z_{s'})$. V has an extension V^* to L such that its maxial ideal is the kernel of a homomorphism over K such that $(z_1, \ldots,z_{s'},b) \rightsquigarrow (r_1,\ldots,r_{s'},b')$. Theorem 5.9.7, (4) shows that the residue class field of V^* is naturally isomorphic to $K(r_1,\ldots,r_{s'}, b)$. QED

COROLLARY 6.5.2 Let K be a field as in Theorem 6.5.1, and let L be a finitely generated field over K. Then there is a K-injection of L into R if and only if (i) L is formally real and (ii) $\text{trans.deg}_K L \leq \text{trans.deg}_K R$.

EXCERCISE 6.1

1.. Assume that a ring R is an ordered set satisfying: (1) $a \geq b$, $c \geq d$ imply $a + c \geq b + d$; (ii) $a > b$ implies $-a < -b$; and (iii) $a > 0$, $b > 0$ imply $ab > 0$. Prove that (1) R is an integral domain, (2) the order is extended to the field K of fractions to make K into an ordered field, and (3) such an extension is unique.

2. Prove that these conditions as above and $a \geq b > 0$, $c \geq d > 0$, imply $ac \geq bd$.

3. Prove that the rational number field has unique structure as an ordered field.

4. An ordered field K is called archimedean if $a, b \in K$, $a > 0$ imply the existence of a natural number n such that $na > b$. Prove that an ordered field is archimedean if and only if it is order-isomorphic to a subfield of the real number field.

5. Prove that if K is formally real, then a pure transcendental extension field over K is also formally real.

6. Prove that if L is an algebraically closed field of characteristic 0, then there is a real closed field K such that $L = K(\sqrt{-1})$.

EXERCISE 6.2

1. Let a be a positive element in an ordered field K. Prove that if $\sqrt{a} \notin K$, then the order can be extended to $K(\sqrt{a})$ exactly in two ways.

2. Prove that the real closure of an archimedean ordered field is archimedean.

3. Let K be a subfield of a formally real field L. Make K into an ordered field. Prove that the order of K can be extended to L (so that L is an ordered field) if and only if $\sum_{i=1}^{n} a_i X_i^2 = 0$ has no nontrivial solution in L whenever all a_i are positive elements in K.

EXERCISE 6.3

1. Let K be a field algebraic over the rational number field. Prove that an element a of K is the sum of certain square elements in K if and only if it holds that: If ψ is an injection of K into the real number field $\mathbb{R}$, then $\psi a > 0$.
2. Assume that a field L contains an ordered field K. Let $a \in L$. Prove that a is expressed in the form $\sum c_i g_i^2$ with $0 < c_i \in K$, $g_i \in L$ if and only if $a > 0$ whenever the order of K is extended to L so that L is an ordered field.

EXERCISE 6.4

1. Under the notation of Theorem 6.4.1, prove that $R = \{x \in K \mid xM \subseteq M\}$.
2. Let S be an ordered subfield of an ordered field T, and assume that for every $t \in T$ there is an $s \in S$ such that $t < s$. Let R be a valuation ring of a field K. Assume that (1) R contains S, (2) there is an S-homomorphism ψ of R into T such that the kernel M is the maximal ideal. Prove that K can be made into an ordered field so that R is the valuation ring of the ordered field K with respect to S, and ψ induces an order-isomorphism of R/M to ψR.

EXERCISE 6.5

1. Let K be a purely transcendental extension of a formally real field K_0. Prove that if a field L is a finite algebraic extension of odd degree over K, then L is formally real.

CHAPTER 7

GALOIS THEORY OF ALGEBRAIC EXTENSIONS OF INFINITE DEGREE

As we defined in Chap.3, the Galois group $G(L/K)$ of a normal extension L of a field K is the group of K-automorphisms $\mathrm{Aut}_K L$. In the finite degree case, the fundamental theorem of Galois states a nice correspondence of subgroups of $G(L/K)$ to intermediate fields. In the infinite degree case, a similar result holds, by defining a natural topology on $G(L/K)$, in the form of a correspondence of closed subgroups to intermediate fields.

7.1 NATURAL TOPOLOGY ON GALOIS GROUPS

LEMMA 7.1.1 Let L, L' be normal extensions of a field K such that $K \subseteq L' \subseteq L$. Then $G(L/L')$ is a normal subgroup of $G(L/K)$ and $G(L/K)/G(L/L') \cong G(L'/K)$. In particular, if $[L' : K] < \infty$, then $\#(G(L/K)/G(L/L')) = [L' : K]_s \leq [L' : K]$.

PROOF: By definition, $G(L/L') \subseteq G(L/K)$. Let ψ be the mapping of $G(L/K)$ to $G(L'/K)$ which maps σ to its restriction $\sigma|_{L'}$. Then the kernel of ψ is $G(L/L')$. If $\sigma \in G(L'/K)$, then σ is extended to an injection of L into a large field; since L is normal, L is mapped to L, and ψ is surjective. QED

Now we define the natural topology on $G = G(L/K)$ by taking $N^* = \{G(L/L') \mid L' \text{ normal over } K,\ [L' : K] < \infty\}$ as a fundamental system of neighborhoods of 1. This definition is justified by:

LEMMA 7.1.2 (1) If $1 \neq \sigma \in G$, then there is $N \in N^*$ such that $\sigma \notin N$. (2) If $N, N' \in N^*$, then $N \cap N' \in N^*$. (3) Let K_λ $(\lambda \in \Lambda)$ be intermediate fields between K and L, and let K^* be the field generated by them. Then $\cap\, G(L/K_\lambda) = G(L/K^*)$. (4) Let $M^* = \{N_\lambda \mid \lambda \in \Lambda\} \subseteq N^*$ be such that (i) $N, N' \in M^*$ implies that $N'' \subseteq N \cap N'$ with $N'' \in M^*$ and (ii) to each $\lambda \in \Lambda$, there is σ_λ such that $N_\lambda \subseteq N_\mu$ $(\lambda, \mu \in \Lambda)$ implies $N_\lambda\sigma_\lambda \subseteq N_\mu\sigma_\mu$. Then there is an element σ of G such that $N_\lambda\sigma_\lambda = N_\lambda\sigma$ for every $\lambda \in \Lambda$.

PROOF: (1): Take $b \in L$ such that $b^\sigma \neq b$, and let L' be the least normal extension containing $K(b)$. Then $\sigma \notin G(L/L') \in N^*$.

(3): Since $G(L/K^*) \subseteq G(L/K_\lambda)$, we have $G(L/K^*) \subseteq \cap_\lambda G(L/K_\lambda)$. Conversely, if $\sigma \in \cap\, G(L/K_\lambda)$, then $a^\sigma = a$ for any $a \in K_\lambda$; hence $a^\sigma = a$ for any $a \in K^*$, and $\sigma \in G(L/K^*)$.

(2): If L', L'' are normal extensions of finite degree over K, then $L' \vee L''$ is also a normal extension of finite degree over K.

(4): Take a normal extension K_λ $(\subseteq L)$ such that $N_\lambda = G(L/K_\lambda)$ for each $N_\lambda \in M^*$. Replacing K_λ with the separable closure of K in K_λ, we may assume that K_λ is a Galois extension of K. Let K^* be the field generated by all of K_λ. We define an automorphism σ of K^* as follows: If $a \in K^*$, then a is contained in a field generated by a finite number of K_λ. By our assumption on M^*, it follows that $a \in K_\lambda$ for a λ. Then we define a^σ to be a^{σ_λ}. Then we see easily that $\sigma \in G(K^*/K)$. σ can be extended to an element of G. QED

Let H be a closed subgroup (i.e., a subgroup which is a closed subset) of a topological group G. Then there are natural mappings $\psi : G \to G/H$, $\phi : G \to H\backslash G$ such that $\psi a = aH$, $\phi a = Ha$. As in the case of residue class groups (cf. Exercise 5.4.1), we introduce a topology on each of G/H, $H\backslash G$ by taking the set of images of open sets as a subbasis of open sets. In this case, G/H or $H\backslash G$ is called a residue class space of G modulo H.

THEOREM 7.1.3 If a topological group G has a family M of closed subgroups enjoying the following five properties, then G is compact.

(1) If $H \in M$, then G/H is compact. (2) If $H, H' \in M$, then

there is $H'' \in M$ such that $H'' \subseteq H \cap H'$. (3) If F is a closed set of G and if $H \in M$, then $FH/H = \{fH \in G/H \mid f \in F\}$ is a closed set in G/H. (4) If U is a neighborhood of 1 in G, then there is $H \in M$ such that $H \subseteq U$. (5) If σ_H $(\in G)$ is associated to each $H \in M$ so that $H \subseteq H'$ implies $\sigma_H H \subseteq \sigma_{H'} H'$, then there is $\sigma \in G$ such that $\sigma_H H = \sigma H$ for all $H \in M$.

PROOF: Consider sets F^* consisting only of closed sets in G and having the finite intersection property. It suffices to show that $\cap_{F \in F^*} F$ is not empty for each F^*. Since the set F^{**} of all F^* is an inductive set, we may consider only maximal members F^* of the set F^{**}. Then F^* enjoys: (i) $F, F' \in F^*$ implies $F \cap F' \in F^*$ and (ii) if F'' is a closed set not belonging to F^*, then $F \cap F''$ is empty for some $F \in F^*$. Now, let H be an arbitrary member of M. Then $\{FH/H \mid F \in F^*\}$ is a set consisting only of closed sets in G/H [by (3) above] and has the finite intersection property. Since G/H is compact, there is σ_H such that $\sigma_H H \subseteq \cap_{F \in F^*} FH/H$. This $\sigma_H H$ is unique, for every member of F^* meets $\sigma_H H$ (as a subset of G); hence by the property (ii) above, we have $\sigma_H H \in F^*$. So, if τH is also in $\cap_{F \in F^*} FH/H$, then $\tau H \in F^*$ and $\sigma_H H$ meets τH, and $\sigma_H H = \tau H$. Similarly, if $H' \in M$, $H' \subseteq H$, then $\sigma_{H'} H' \in F$ and therefore $\sigma_{H'} H' \subseteq \sigma_H H$. Therefore by (5) above, there is $\sigma \in G$ such that $\sigma H = \sigma_H H$ for all $H \in M$. If $\sigma \notin F$, then there is a neighborhood U of 1 such that σU does not meet F. Then, by an $H \in M$ such that $H \subseteq U$, we see that σH does not meet F, a contradiction. Thus $\sigma \in F$. QED

Now we come bake to $G = G(L/K)$, and we have:

THEOREM 7.1.4 G is a compact topological group. If K' is an intermediate field, then $G(L/K')$ is a closed subgroup of G. If furthermore $[K' : K]$ is finite, then $G(L/K')$ is open and closed.

PROOF: We apply Theorem 7.1.3 to our G with $M = N^*$ in the notation of Lemma 7.1.2. Then we see that G is compact. If $[K' : K]$ is finite, then the least normal extension K'' containing K' over K is of finite degree. Since $G(L/K'')$ is a member of N^*, $G(L/K'')$ is an open set. Then $G(L/K')$ is the union of open sets

$\sigma(G(L/K''))$ $[\sigma \in G(L/K')]$, and therefore $G(L/K')$ is an open set. Then Theorem 5.4.5 shows that $G(L/K')$ is closed. In the general case, K' is generated by all of its subextensions K'' of finite degrees over K, and therefore $G(L/K')$ is the intersection of these $G(L/K'')$ [Lemma 7.1.2, (3)]. Since $G(L/K'')$ are closed, we see that $G(L/K')$ is closed. QED

THEOREM 7.1.5 Under the notation as above, the natural topology of $G(L/K')$ coincides with its topology as a subspace of $G(L/K)$.

PROOF: If K'' is a normal extension of K and $[K'' : K] < \infty$, then $K' \vee K''$ is a normal extension of K' of a finite degree. Conversely, a normal extension of K' of finite degree is contained in such a $K' \vee K''$. This and the definition of the topology proves the assertion. QED

7.2 THE FUNDAMENTAL THEOREM OF GALOIS

Let L be a Galois extension of a field K (i.e., L is separable and normal over K), and we consider $G = G(L/K)$ with its natural topology. Then the fundamental theorem of Galois in the general case is stated as follows:

THEOREM 7.2.1 There is a one-one correspondence between the set H^* of closed subgroups H of G and the set M^* of intermediate fields M in such a way that if H corresponds to M, then (i) $H = G(L/M)$ and (ii) $M = \{x \in L \mid x^\sigma = x$ for every $\sigma \in H\}$.

In this case, (1) H is a normal subgroup of G if and only if M is a Galois extension of K; (2) if each H_λ $(\lambda \in \Lambda)$ corresponds to M_λ, then $\cap H_\lambda$ corresponds to the field generated by all the M_λ and $\cap M_\lambda$ corresponds to the closure of the subgroup generated by all the H_λ.

PROOF: For each $H \in H^*$, we set $M(H) = \{x \in L \mid x^\sigma = x$ for any $\sigma \in H\}$. $M(H)$ is an intermediate field, and $G(L/M(H)) = H$. [Indeed, obviously $H \subseteq G(L/M(H))$. If $\sigma \notin H$, then there is a neighborhood U of 1 such that σU does not meet H. We may assume that $U = G(L/K')$ with a Galois extension K' of a finite degree over K. Then $\sigma U \notin HU/U$. HU/U corresponds to $M(H) \cap K'$ by the finite degree case. The fact that $\sigma U \notin HU/U$ implies that there is an element b in $M(H) \cap K'$ such that $b^\sigma \neq b$ and $\sigma \notin G(L/M(H))$.] Conversely, assume that $M \in M^*$ and $H = G(L/M)$. Then $H \in H^*$ by Theorem 7.1.4. Obviously $M \subseteq M(H)$. If $M(H) \neq M$, then there is an element a of $M(H)$ which is not in M. Since a is separable over K, there is an element σ of $G(L/M)$ such that $a^\sigma \neq a$, which contradicts $a \in M(H)$ because $H = G(L/M)$. Thus (i), (ii) are proved. (1) follows easily from the fact that $H = G(L/M)$, $\sigma \in G$ imply $G(L/M^\sigma) = \sigma^{-1}H\sigma$. The first half of (2) follows from Lemma 7.1.2. As for the last half of (2), set $H = G(L/\cap_\lambda M_\lambda)$. Since $H \supseteq H_\lambda$, H contains the subgroup H' generated by all the H_λ. The closure H'' of H' is a subgroup as will be shown below (Lemma 7.2.2). Let M'' be the intermediate field corresponding to H''. Then $M'' \subseteq M_\lambda$ and $M'' \subseteq \cap_\lambda M_\lambda$. On the other hand, $H \supseteq H'$ and H is closed; hence $H \supseteq H''$, which shows that $\cap M_\lambda \subseteq M''$. QED

LEMMA 7.2.2 If H is a subgroup of a topological group G, then the closure H'' of H is a subgroup of G.

PROOF: Assume that $a,b \in H''$, and it suffices to show $ab^{-1} \in H''$. Let V be an arbitrary neighborhood of ab^{-1}. Then by the continuity of multiplication, there is a neighborhood U of 1 such that $UaUb^{-1} \subseteq V$. bU^{-1} is a neighborhood of b. Since $a,b \in H''$, there are a', b' such that $a' \in H \cap Ua$, $b' \in H \cap bU^{-1}$. Then $a'b'^{-1}$ is in both V and H. Thus $V \cap H$ is not empty and $ab^{-1} \in H''$. QED

THEOREM 7.2.3 Let L be a normal extension of a field K, and set $G = G(L/K)$. For an intermediate field M, we set $H = \{\sigma \in G \mid x^\sigma = x$ for any $x \in M\}$ and $M' = \{a \in L \mid a^\sigma = a$ for any $\sigma \in H\}$. Then $H = G(L/M) = G(L/M')$, M' is purely inseparable over M, and L is separable over M'.

PROOF: Obviously $M \subseteq M'$, $H = G(L/M) = G(L/M')$. If a $(\in M')$ is not purely inseparable over M, then there is an injection σ of $M(a)$ to L over M such that $a^\sigma \neq a$ and σ is extended to an element of $G(L/M)$. This contradicts $a \in M'$. QED

7.3 SPLITTING FIELDS, INERTIA FIELDS, RAMIFICATION FIELDS

Assume that R is a normal integral domain, K its field of fractions, and K^* a normal extension of K with Galois group $G = G(K^*/K)$. Let R^* be the integral closure of R in K^*. Let P^* be a prime ideal of R^*. As we defined in Sec.5.11, $H_{P^*} = \{\sigma \in G \mid P^{*\sigma} = P^*\}$ is the splitting group of P^*, and $S_{P^*} = \{x \in R^* \mid x^\sigma = x$ for any $\sigma \in H_{P^*}\}$ is the splitting ring of P^*. For an ideal I^* of R^*, $V_{I^*} = \{\sigma \in G \mid x^\sigma - x \in I^*$ for any $x \in R^*\}$ is called the <u>ramification group</u> defined by I^*. A <u>ramification group</u> of P^* is a V_{I^*} with P^*-primary I^*. Especially, V_{P^*} is called the <u>inertia group</u> of P^*. $T_{I^*} = \{x \in R^* \mid x^\sigma = x$ for any $\sigma \in V_{I^*}\}$ is called the <u>ramification ring</u> defined by I^*. If I^* is P^*-primary, then T_{I^*} is a <u>ramification ring</u> of P^*, and T_{P^*} is the <u>inertia ring</u> of P^*. Fields of fractions of S_{P^*}, T_{I^*}, and T_{P^*} are called the <u>splitting field</u> of P^*, the <u>ramification field</u> defined by I^* (or a <u>ramification field</u> of P^* in case I^* is P^*-primary), and the <u>inertia field</u> of P^*, respectively.

THEOREM 7.3.1 These H_{P^*} and V_{I^*} are closed subgroups of G.

PROOF: Assume that $\sigma \notin H = H_{P^*}$ ($\sigma \in G$). Then $P^{*\sigma} \neq P^*$, and there is b in P^* such that $b^\sigma \notin P^*$. Let K' be the least normal extension containing b over K, and set $N = G(K^*/K')$. Then $N\sigma$ is a neighborhood of σ. If $\tau \in N$, then $b^{\tau\sigma} = b^\sigma \notin P^*$, which means that $N\sigma$ does not meet H. Thus H is closed. Next, consider $V = V_{I^*}$. Assume that $\sigma \notin V$ ($\sigma \in G$). Then there is c in R^* such that $c^\sigma - c \notin I^*$. Taking the least normal extension K' containing c, we set $N = G(K^*/K')$. Then $N\sigma$ is a neighborhood of σ, and for any $\tau \in N$, we have $c^{\tau\sigma} - c = c^\sigma - c \notin I^*$ and $N\sigma \cap V$ is empty. QED

A generalization of Theorem 5.11.9 is given in the following form:

Under the notation as above, assume furthermore that K^* is separable over K. Set $P = P^* \cap R$, $S = S_{P^*}$, $Q = P^* \cap S$, $T = T_{P^*}$, $Q' = P^* \cap T$, $L = R_P/PR_P$, $L' = T_{Q'}/PT_{Q'}$, and $L^* = R^*_{P^*}/P^*R^*_{P^*}$. Then

THEOREM 7.3.2 (1) P^* is the unique prime ideal of R^* lying over Q; (2) $S_Q/QS_Q = L$; (3) $QS_Q = PS_Q$; (4) $Q'T_{Q'} = PT_{Q'}$; (5) $V = V_{P^*}$ is a normal subgroup of $H = H_{P^*}$; (6) L^* is a normal extension of L, H/V is naturally identified with $G(L^*/L)$; (7) L^* is purely inseparable over L', L' is a Galois extension of L and $G(L'/L) = G(L^*/L)$.

PROOF: Considering the rings of quotients with respect to $R - P$, we may assume that P is the unique maximal ideal of R. $H = G(R^*/S)$ by Theorem 7.3.1, and we see (1) by Theorem 4.7.7. As for (5), let σ, τ be arbitrary elements of V, H, respectively. If $x \in P^*$, then $x^\sigma - x \in P^*$ and $x^\sigma \in P^*$, which implies $V \subseteq H$. For $y \in R^*$, set $z = y^{\tau^{-1}\sigma\tau} - y = (z^\sigma - z)^\tau \in P^{*\tau} = P^*$. Thus (5) is proved. (2), (3), (4), (6), (7) are easily reduced to the case where $[L^* : K] < \infty$, and therefore (2), (3) follow from Theorem 5.11.9. In order to prove (4), (6), (7), we assume that G is finite. Since $L = S_Q/QS_Q$, we may assume furthermore that $R = S$, i.e., $G = H$. For an arbitrary element a^* of L^*, let $a \in R^*$ be such that $a^* = (a \text{ modulo } P^*)$, and let $f(X)$ be the minimal polynomial for a over R. Then a^* is a root of $f^*(X) = (f(X) \text{ modulo } P)$. Since $f(X)$ is the product of linear factors in $R^*[X]$, we see that every conjugate of a^* is of the form $(a^\sigma \text{ modulo } P^*)$ with $\sigma \in G = H$. This proves that L^* is normal over L. If this observation is applied to R^* over T, we see that L^* is purely inseparable over L' in view of the fact that elements of V fix every a^*. Thus, in view of the fact that V is a normal subgroup of $H = G$, we may assume that $R^* = T$, i.e., $V = \{1\}$. Let $1, b, \ldots, b^{m-1}$ ($b \in T = R^*$) be such that their residue classes b^{*i} modulo Q' ($= P^*$) form a linearly independent base for a maximal simple subextension of $L^* = L'$ over L. Set $T'' = \sum_{i=0}^{m-1} Rb^i$. Since $V = \{1\}$, every element σ ($\neq 1$) of G induces a nontrivial automorphism of L^*, and therefore $m \geq \#(G)$. Since b is integral over the normal ring R, $1, b, \ldots, b^{m-1}$ must be linearly independent over R, and we have $m \leq \#(G)$. This proves that L^* is a Galois extension of L and that $G(L^*/L) = G$. Thus we have proved (6) and (7). Furthermore, letting

$g(X)$ be the minimal polynomial for b over R, we see that $\deg g(X) = m$ and that ($g(X)$ modulo P) has no multiple root and is the minimal polynomial for b^* over L; hence the discriminant d of $g(X)$ is a unit in R. Thus $R^* = R[b]$ (Theorem 4.10.2), and we see (4) easily. QED

With notation as before, we observe V_{I^*}. The first remark is:

LEMMA 7.3.3 Let U be a subset of R consisting only of non-zero-divisors modulo I^* (i.e., $a \in R^*$, $b \in U$, $ab \in I^*$ imply $a \in I^*$). Let V_I be the ramification group of $I = I^*R_U^*$ with respect to the normal extension R_U^* over R_U. Then V_{I^*} coincides with V_I.

PROOF: $I \cap R^* = I^*$, and the coincidence follows easily. QED

Our next remark is applicable to the case where $P^*R_{P^*}$ is a principal ideal and $R_{P^*}/P^*R_{P^*}$ is perfect.

THEOREM 7.3.4 Assume furthermore that $P^* = pR^*$ ($p \in R^*$) and that the field $L = R_P/PR_P$ is perfect. Denoting $V_{p^nR^*}$ by V_n ($n = 1, 2, \ldots$), we see that each V_{n+1} is a normal subgroup of V_n and V_n/V_{n+1} is isomorphic to an additive subgroup of R^*/P^*.

PROOF: Let R_i be the ring of V_i-invariants in R^* for each i. Then $R \subseteq R_1 \subseteq R_2 \subseteq \cdots \subseteq R_n \subseteq \cdots \subseteq R^*$ obviously. Considering the inertia ring R_1 instead of R, we may assume that $R = R_1$. Then Theorem 7.3.2 shows that $L^* = R_{P^*}^*/pR^*_{P^*}$ coincides with L. Let $\sigma \in V_n$. Then $p^\sigma - p \in p^nR^*$, and we can write $p^\sigma = p + p^nc(\sigma)$ [$c(\sigma) \in R^*$]. Thus, for each $\sigma \in V_n$, we can associate $a(\sigma) = (c(\sigma)$ modulo $P^*)$. If $\sigma, \tau \in V_n$, then $p^{\sigma\tau} = (p + p^nc(\sigma))^\tau = p + p^nc(\tau) + (p+p^nc(\tau))^nc(\sigma)^\tau$. Since $\tau \in V_n \subseteq V_1$, $c(\sigma)^\tau - c(\sigma) \in pR^*$, and therefore $a(\sigma\tau) = a(\sigma) + a(\tau)$. If $\sigma \in V_{n+1}$, then $a(\sigma) = 0$. Conversely, if $a(\sigma) = 0$, then $p^\sigma - p \in p^{n+1}R^*$. Let $x \in R^*$. Since $L^* = L$, there is d in R such that $d \notin pR^*$ and (dx modulo pR^*) $\in R/P$, and we can write $dx = h + pk$ ($h \in R$, $k \in R^*$). Then $dx^\sigma = (dx)^\sigma = h + p^\sigma k^\sigma$, $p^\sigma - p \in p^{n+1}R^*$, $k^\sigma - k \in p^nR^*$ (because $\sigma \in V_n$), and we see that $dx^\sigma - dx \in p^{n+1}R^*$ and $x^\sigma - x \in p^{n+1}R^*$. Thus V_{n+1} is the kernel of the homomorphism $\sigma \rightsquigarrow a(\sigma)$. QED

EXERCISE 7.1

1. Let L, L' be normal extensions of a field K, and K' a field such that $K \subseteq K' \subseteq L' \subseteq L$. Prove that the following three are equivalent to each other: (1) $G(L'/K') = \{1\}$. (2) L' is purely inseparable over K'. (3) $G(L/L') = G(L/K')$.
2. Prove that the natural isomorphism in Lemma 7.1.1 is an isomorphism of topological groups.
3. Let F^* be the algebraic closure of a finite field F.
   (1) Prove that the additive group Z of rational integers becomes a topological group by taking $N^* = \{nZ \mid n = 1, 2, 3, \ldots\}$ as a fundamental system of neighborhoods of 0.
   (2) Prove that there is an injection ψ of the topological group Z [given in (1) above] to the Galois group $G(F^*/F)$ such that (i) ψZ is dense in $G(F^*/F)$ and (ii) with $q = \#(F)$, it holds that $a^{\psi 1} = a^q$ for every $a \in F^*$.
   (3) Prove that $G(F^*/F)$ is commutative, that $G(F^*/F)$ has no elements of finite order except 1, and that the topological group structure of $G(F^*/F)$ is independent of the choice of F.

EXERCISE 7.2

1. Let K^* be the algebraic closure of a field K, and let H be the set of elements σ of $G(K^*/K)$ such that $\sigma \neq 1$ and σ is of finite order. Prove that (i) there is a one-one correspondence between H and the set of real closed fields over which K^* is algebraic; (ii) if $\sigma \in H$, then $\sigma^2 = 1$; and (iii) H is not empty if and only if K is formally real.
2. Let P be the prime field of a field K, and let L be a normal extension of K. Prove the following three assertions: (i) $\mathrm{Aut}_P\, L$ becomes a topological group by taking a basis of open sets to be the set Ω of all $v(a,b)$ obtained in the following manner: n is an arbitrary natural number, $(a) = (a_1, \ldots, a_n)$, $(b) = (b_1, \ldots, b_n)$ $(a_i, b_j \in L)$ and $v(a,b) = \{\sigma \in \mathrm{Aut}_P\, L \mid a_i^{\sigma} = b_i$ for

$i = 1, \ldots, n\}$. (ii) The Galois group $G(L/K)$ is a subspace of $\mathrm{Aut}_P\, L$ [given in (i) above]. (iii) If H is a compact subgroup of $\mathrm{Aut}_P\, L$, then $H = G(L/M)$ with the field M of H-invariants in L.

EXERCISE 7.3

1. Under the notation as in Sec.7.3, we may define the <u>splitting group</u> of the ideal I^* to be $H_{I^*} = \{\sigma \in G \mid I^{*\sigma} = I^*\}$. Prove that H_{I^*} is a closed subgroup of $G(K^*/K)$. Prove an adaption of Lemma 7.3.3 to H_{I^*}.
2. Under the notation as above, let P^{**} be another prime ideal of R^*. (1) Show by examples that (i), (ii) below are not correct. (2) Prove that if another ideal I^{**} of R^* is contained in I^*, then $V_{I^{**}} \subseteq V_{I^*} \subseteq H_{I^*}$. (3) Prove that if R is a valuation ring, then (ii) holds grood:

   (i) If $P^{**} \subseteq P^*$, then $H_{P^{**}} \subseteq H_{P^*}$.

   (ii) If $P^{**} \subseteq P^*$, then $H_{P^*} \subseteq H_{P^{**}}$.

ANSWERS AND COMMENTS ON EXERCISES

CHAPTER 1

EXERCISE 1.1

1. Z/nZ.
2. If $1 \neq a \in G$, then $\#(\langle a \rangle)$ $(\neq 1)$ is a divisor of $\#(G)$.
3. The symmetric group of degree 3.
5. Let a be a generator of G. $(a^m)^d = 1$ if and only if n divides md, namely, m is a multiple of n/d. Thus $H = \langle a^{n/d} \rangle$.

EXERCISE 1.2

3. Let $G = \langle a \rangle$, $H = \langle b \rangle$. f $(\in K)$ is uniquely determined by s such that $fa = b^s$ $(0 \leq s < h)$. The condition is that $(b^s)^g = 1$, namely, s is a multiple of h/d. Thus $\#(K) = d$.
4. If $a \notin H$ $(a \in G)$, then $aH = G - H = Ha$.

EXERCISE 1.3

3. Example 1. Consider the symmetric group G on the set $\{1,2,3,4\}$. Using the notation in Sec.2.8, $H = \{1, (1,2)(3,4), (1,3)(2,4), (1,4)(2,3)\}$ is a normal subgroup of G, and $K = \{1, (1,2)(3,4)\}$ is a normal subgroup of H. But K is not a normal subgroup of G.

   Example 2. Let G be the set of matrices $\begin{pmatrix} a & b \\ 0 & c \end{pmatrix}$ with rational numbers a, b, c. Let H be the set of those matrices such that $a = c = 1$, and let K be the set of those such that $a = c = 1$, $b \in Z$.
4. The difinition of solvability implies that if N is a subgroup of a solvable group G, then N is solvable. Exercise 1.3.2 shows that homomorphic images of a solvable group are solvable.

EXERCISE 1.4

1. $a^{-1}Sa = b^{-1}Sb$ if and only if $ba^{-1} \in N$.
2. If $b \in N(K)$, then $b^{-1}Hb \subseteq K$, and therefore $b^{-1}Hb = c^{-1}Hc$ with $c \in K$. Then $bc^{-1} \in N(H) \cap N(K)$.
3. Apply Exercise 1.4.2.

EXERCISE 1.6

1. If $a \in H$, $b \in K$, then the order of (a,b) in $H \times K$ is the least common multiple of the orders of a and b.
2. For each natural number n, let $\phi(n)$ be $\#(\{m \in \mathbb{N} \mid m \leq n,\ m$ has no proper common factor with $n\})$ (Euler function; cf. Sec. 3.8). Let $D(n)$ be the set of natural numbers which divide n. Then we have $n = \sum_{d \in D(n)} \phi(d)$. Now, if m is the order of an element a of G, then $m \in D(\#(G))$. Furthermore, our assumption implies that if the order of an element b divides m, then $b \in \langle a \rangle$. Therefore the number of elements of order m in G coincides with $\phi(m)$. Since the sum of them must be $\#(G)$, there must be an element of order $\#(G)$.

CHAPTER 2

EXERCISE 2.3

2. No. Indeed, take $e_i \in I_i$ so that $1 = e_1 + \cdots + e_n$. For $i \neq j$, $I_i \cap I_j = 0$ and $I_iI_j = 0$. Therefore, for $x \in I_j$, $x = \sum xe_i = xe_j$. In particular, $e_j = e_j^2$, $I_j = e_jR$. Therefore R is the direct sum of $I_1, \ldots, I_n$.

EXERCISE 2.5

1. If $a \neq 0$, $a^2 = 0$, then $(aX + 1)(aX - 1) = 1$. Conversely, assume that $fg = 1$ $(f,g \in R[X])$. Then Corollary 2.5.2 shows that if P is a prime ideal of R, then the coefficients of positive degree terms of f and g are in P; hence they are nilpotent by the first half of Theorem 4.6.4 (whose proof is easily understood by virtue of results in Secs. 2.4 and 2.5).
2. If $ab = 0$, then $(aX + 1)(bX + 1) = (a + b)X + 1$. Note that it can happen that $\deg fg < \deg f + \deg g$ if and only if R is not an integral domain.

EXERCISE 2.6

1. If part: It suffices to show that a proper irreducible element p is a prime element. Assume that $ab \in pR$, $a \notin pR$. Then ab is a common multiple of a and p, and hence ab is a multiple of ap. Thus b is a multiple of p.

EXERCISE 2.7

2. If a, b are rational numbers, then they are linearly dependent.
4. The first half: $\{m + m \mid m \in M\}$ in $M \oplus M$. The last half: Let e, e' be the identities of R, R'. Then $A = Ae \oplus Ae'$.

EXERCISE 2.8

1. The symmetric group is $A_n \cup \tau A_n$. Therefore $\tau A_n = A_n \tau$. Thus $f - \tau f$ is an alternating form or 0; $f + \tau f$ is a symmetric form.

EXERCISE 2.9

1. If $f(X) = g(X)h(X)$, then, since $f(X) \equiv c_0 X^n$ modulo P, we see that $g(X) \equiv aX^m$ modulo P, $h(X) \equiv bX^{n-m}$ modulo P. If $m > 0$, $n - m > 0$, then the constant term of $f(X)$ is in P^2.

CHAPTER 3

EXERCISE 3.1

2. (1) 4, (2) 2, (3) 3, (4) 6.

EXERCISE 3.2

1. (i) $Q(\sqrt{5}, \sqrt{-3})$; degree 4. In finding the roots, note that $X^4 - X^2 + 4 = (X^2 + 2)^2 - 5X^2$. (ii) $Q(\sqrt[3]{2}, \sqrt{-3})$; degree 6. (iii) $Q(\sqrt{-1}, \sqrt{2})$; degree 4.

EXERCISE 3.3

1. (i) Characteristic 2 (1 is a 4-ple root). (ii) Characteristic 229 (-4/3 is a double root). (iii) Characteristic 2 (primitive cubic roots of unity are double roots), or 3 (1 is a 4-ple root) or 139 (-7 is a double root), (iv) This has a double root (actually p-ple or more) if and only if $c_n = 0$.
3. Show that $t^{1/p} \notin K(t)$.

5. (iii) Note that $K^{p^{-\infty}}$ is perfect; hence this is isomorphic to K if and only if K is perfect.

6. a is purely inseparable over $K(a^p)$. Therefore, if a is separable over K, then $a \in K(a^p)$. If a is inseparable over K, then $[K(a^p) : K] < [K(a) : K]$.

EXERCISE 3.4

1. $K(x,y)$ over $K(x^p,y^p)$, provided that x, y are algebraically independent over a field K of characteristic $p \neq 0$.

2. Let $L = K(a)$, and let $f(X)$ be the minimal polynomial for a over K. As was shown in the proof of Theorem 3.4.5, an intermediate field M corresponds to a factor $f_M(X)$ of $f(X)$ having a as a root. The number of such factors is at most 2^{n-1}.

EXERCISE 3.5

1. First half: Conjugates of a separable elements are separable. The last half: Let x, y be algebraically independent elements over a field K_0 of characteristic $p \neq 0$. Set $K = K_0(x + y^p, xy^p)$, $L = K(x,y)$ $(= K_0(x,y))$.

3. $L = Q(\sqrt{2})$, $L' = Q(\sqrt[4]{2})$, $K = Q$.

EXERCISE 3.6

1. (i) $\{1\}$. (ii) The group is generated by σ, τ such that $(\sqrt{-1})^\sigma = \sqrt{-1}$, $(\sqrt[4]{2})^\sigma = \sqrt{-1}\cdot\sqrt[4]{2}$, $(\sqrt{-1})^\tau = -\sqrt{-1}$, $(\sqrt[4]{2})^\tau = \sqrt[4]{2}$. Then $\sigma^4 = 1$, $\tau^2 = 1$, $\tau\sigma\tau = \sigma^{-1}$. (iii) If K is of characteristic 2, then the group is $\{1\}$; otherwise the group is $\{1,\sigma\}$, $t^\sigma = -t$. (iv) If K contains a cubic root ω of unity other than 1, then the group is $\{1,\sigma,\sigma^2\}$, $t^\sigma = \omega t$. Otherwise the group is $\{1\}$.

2. (i) $Q(\sqrt[3]{2})$. (ii) Q. (iii) $K(t)$ if the characteristic of K is 2; K otherwise. (iv) K if $\omega \in K$; otherwise $K(t)$.

3. Note first that $Tr_{M/K}$ and $N_{M/K}$ are independent of the choice of L, and (i), (ii) are easy. As for (iii), prove first that the determinant and the trace of $\rho(a)$ are independent of the choice of a linearly independent base and then proceed as follows: Let $d_1 = 1, d_2, \ldots, d_v$ be a linearly independent base of M

over $K(a)$, and set $r = [K(a) ; K]$ $(n = rv)$. Let $b_1, \ldots, b_n$ be the $d_i a^j$ $(i = 1, \ldots, v;\ j = 0, 1, \ldots, r - 1)$ and compute the trace and the determinant of $\rho(a)$. As for (v), the if part is obvious and we assume that M is separable over K. (If one knows Lemma 3.8.9 already, then the assertion follows from the Lemma easily.) If $[M : K]$ is not a multiple of the characteristic p of K, then $Tr_{M/K}\, b \neq 0$ for $0 \neq b \in K$. Therefore we assume that $[M : K]$ is a multiple of p $(\neq 0)$. Take $a \in M$ such that $M = K(a)$ and let $f(X) = X^n + c_1X^{n-1} + \cdots + c_n$ be the minimal polynomial for a over K. Let $a_1 = a, a_2, \ldots, a_n$ be the conjugates of a and let j be such that $c_j \neq 0$, $c_i = 0$ for any $i < j$. Set $m_i = \sum_k a_k{}^i$. Then we have $m_i = 0$ for $i < j$ and $m_j \neq 0$.

EXERCISE 3.7

1. The first: The Galois group is isomorphic to the symmetric group of degree 3, and it has (i) exactly one subgroup of order 3, (ii) 3 of order 2. Answer: $1 + 1 + 3 + 1 = 6$. The last: The Galois group G of $Q(\sqrt{-1}, \sqrt[4]{2})$ is generated by σ, τ as was shown above for Exercise 3.6.1 (ii), and $Q(\sqrt[4]{2})$ corresponds to its subgroup $H = \{1,\tau\}$. Thus intermediate fields correspond to subgroups of G containing H, and these subgroups are H, $\{1,\tau, \sigma^2, \tau\sigma^2\}$, G.

EXERCISE 3.8

2. Cf. Corollary 3.8.4 and Theorem 3.8.5. If p is an odd prime number, then the unit group U_n of Z/p^nZ is a cyclic group, as can be proved by induction on n.
4. Let a be such that $L = K(a)$. Then $1, a, \ldots, a^{n-1}$ form a linearly independent base. Apply the following equality to this case:

$$\det \begin{pmatrix} 1 & X_1 & X_1{}^2 & \cdots & X_1{}^{n-1} \\ 1 & X_2 & X_2{}^2 & \cdots & X_2{}^{n-1} \\ \cdot & \cdot & \cdot & \cdot & \cdot \\ 1 & X_n & X_n{}^2 & \cdots & X_n{}^{n-1} \end{pmatrix} = \Pi_{i>j}\,(X_i - X_j)$$

6. (ii) Use Exercise 3.8.4 for the only if part. Use Exercise 3.8.5 for the existence.

7. (i) Only if part: Set $b_i = aa^{\sigma}\cdots a^{\sigma^{i-1}}$ $(i = 1, \ldots, n = [L : K])$. $b_i^{\sigma} = a^{-1}b_{i+1}$ $(i < n)$, $b_n = 1 = b_n^{\sigma} = a^{-1}b_1$. By Lemma 3.8.9, there is $u \in L$ such that $y = \sum b_i u^{\sigma^i} \neq 0$. Then $y^{\sigma} = ya^{-1}$ and $a = y^{1-\sigma}$. (ii) Only if part: Set $c_i = a + a^{\sigma} + \cdots + a^{\sigma^{i-1}}$ $(i = 1, \ldots, n)$. There is $u \in L$ such that $t = \sum u^{\sigma^i} \neq 0$. Set $z = \sum t^{-1}c_i u^{\sigma^i}$. Then $z - z^{\sigma} = \sum t^{-1}cu^{\sigma^i} - \sum t^{-1}(c_{i+1} - a)u^{\sigma^{i+1}} = a$.

EXERCISE 3.9

1. Use Exercise 1.3.1.

3. (i) and (ii): Discriminants are not square elements, and therefore the Galois groups cannot be a cyclic group of order 3. Thus we see that the Galois group in each case is the symmetric group of degree 3. (iii) The cyclic group of order 2. ($Q(\sqrt{-3})$ is the minimal splitting field.) (iv) Noncyclic group of order 4. ($Q(\sqrt{-1}, \sqrt{-3})$ is the minimal splitting field.) (v) If $n = 2$, then the group is a cyclic group of order 2. Assume that $n \geq 3$. Let Z_n be a cylic group of order n, and let U_n be the unit group of Z/nZ. Let G be the group generated by Z_n and U_n with defining relation that if $a \in U_n$, $b \in Z_n$ and if a' $(\in Z)$ is a representative of a, then $a^{-1}ba = b^{a'}$. Then the Galois group is isomorphic to G.

EXERCISE 3.10

1. By Theorem 3.10.5, we see that $1°$ is not constructed and $3°$ is constructed.

EXERCISE 3.11

3. Fix a natural number $n > 1$, and let K_i be the finite field such that $\#(K_i) = p^{n^i}$ (p is a fixed prime number). Set $K = \cup_i K_i$.

CHAPTER 4

EXERCISE 4.1

3. Let T be a transcendence base of Ω over K and assume that $\#(T)$ is infinite. Let T' be a proper subset of T such that $\#(T) = \#(T')$. Let f be a one-one correspondence between T and T'. Let L and L' be the fields generated by T and T' over K, respectively. Then f is extended to an isomorphism of L to L'. But the isomorphism cannot be extended to an automorphism of Ω.

EXERCISE 4.3

1. Note first that if a field L has a p-base B consisting of a finite number of elements over a subfield K, then in view of Theorem 4.3.7, every p-base of L over K consists of the same number of elements as B. If $L = K(a_1,\ldots,a_n)$, then there is a p-base which is a subset of $\{a_1,\ldots,a_n\}$.
2. One should note here that, letting D' be an extension of D to L, the restriction of D' to a given intermediate field M need not be an extension of D to M because $D'(M)$ may not be contained in M.

   If M has a separating transcendence base, then D is extended to M, and we assume that the characteristic p is different from 0. Take a p-base B of L (over L^p) so that $L^p \vee K = L^p(K \cap B)$, $L^p \vee M = L^p(M \cap B)$. Then extend D so that, for each $x \in (M \cap B) - (K \cap B)$, Dx is in M.

EXERCISE 4.4

2. One can reduce easily to the finitely generated case. In that case, we can use separating transcendence bases.
3. $[M : K] \leq i(L/K)$ is easily verified by observing $L \otimes_K M$. As for the other inequality, use Theorem 4.4.3 and the fact that $[L \vee K^{p^{-\infty}} : K^{p^{-\infty}}(z_1,\ldots,z_n)] \geq [L : K(z_1,\ldots,z_n)]_s$; this last can be proved using Theorem 3.6.3, (2).

EXERCISE 4.5

3. Use Corollary 4.4.5 and Theorem 4.5.2

EXERCISE 4.6

2. Example: Let x, y be algebraically independent elements over a field K, and set $z_n = y/x^n$ $(n = 1, 2, \ldots)$. With $R = K[x, y, z_1, z_2, \ldots]$ and $I = xR$, we have $R/I \cong K$ and R is not noetherian.

   The case where $I^n = 0$: That R/I^m is noetherian is proved by induction on m; either directly or using Exercise 4.6.1.

4. M/I is the unique maximal ideal of R/I.

5. If $bc \in Q : a$, $b \notin Q : a$, then $abc \in Q$, $ab \notin Q$.

6. Let $I = Q_1 \cap \cdots \cap Q_n$ be the shortest representation of I as an intersection of primary ideals. Then $I : a = \cap (Q_i : a)$, and the if part is easy. Only if part: We may assume that $P = \sqrt{Q_1}$. Take c in $Q_2 \cap \cdots \cap Q_n$ outside of Q_1. Then $I : c = Q_1 : c$, which is P-primary (Exercise 4.6.5). Taking d in $(Q_1 : c) : P$ outside of $Q_1 : c$, we may set $a = cd$.

7. (ii): $R_i \subseteq R_0[f_1, \ldots, f_N, \ldots]$ is proved by induction on i. (iii) is proved by using (ii). (iv), if part: It suffices to show that $f_i \in R_i$, $g_j \in R_j$, $(\sum_{i=u}^{u+s} f_i)(\sum_{j=v}^{v+t} g_j) \in Q$, $\sum f_i \notin P$ imply $\sum g_i \in Q$. We use a double induction on s and t. The case $s = 0$ is easy; similar is the case $t = 0$. Assume that $st > 0$. Since $f_u g_v \in Q$, we see that $(\sum f_i)(\sum_{j=v+1}^{v+t} f_u g_j) \in Q$. By induction on t, we have $\sum f_u g_j \in Q$ and each $f_u g_j \in Q$. So (1) if $f_u \notin P$, then each $g_j \in Q$. (2) On the other hand, since $f_u g_j \in Q$, we have $(\sum_{i=u+1}^{u+s} f_i)(\sum g_i) \in Q$; therefore if $f_u \in P$, we use induction on s. (v): Adapt our proof of Theorems 4.6.6, 4.6.9 to homogeneous ideals.

EXERCISE 4.8

1. If a prime ideal P of R contains I, then there is a prime ideal P' of R' lying over P and $\text{ht } P' = \text{ht } P$ (Theorem 4.8.2). From this we see that $\text{ht } I \geq \text{ht } IR'$. Since $IR' \cap R$ contains I, we have $\text{ht } I \leq \text{ht } (IR' \cap R) = \text{ht } IR'$.

EXERCISE 4.9

2. The last half: Let x_i be (X_i modulo P), and set $L_i = K(x_1, \ldots, x_i)$ $(K = L_0)$. Let $f_i(X_1, \ldots, X_i)$ $(i = 1, \ldots, n)$ be a

monic polynomial in X_i such that $f_i(x_1,\ldots,x_{i-1},X)$ is the minimal polynomial for x_i over L_{i-1}. Then P is generated by these f_i. The first half: Let $z_1, \ldots, z_{n-h}$ be elements of $K[X_1,\ldots,X_n]$ such that their residue classes modulo P are algebraically independent over K and such that $z_1, \ldots, z_{n-h}$, $X_1, \ldots, X_h$ generate $K[X_1,\ldots,X_n]$ over K. Let S be the set of nonzero elements in $K[z_1,\ldots,z_{n-h}]$. Then $K[X_1,\ldots,X_n]_S = K(z_1,\ldots,z_{n-h})[X_1,\ldots,X_h]$, and we can apply the last half to $PK[X_1,\ldots,X_n]_S$.

EXERCISE 4.10

1. (1): $R[b] \subseteq a^{-1}R$ and $R[b]$ is a finite module over R.
3. Note that the assertion becomes false if we omit the assumption that R is finitely generated (see, for example, Nagata, Local Rings, John Wiley, New York, reprint ed., Krieger, Huntington, N.Y., pp. 204-205). The required adaption can be proved by adapting our proof of Theorem 4.8.2, and by virtue of Corollary 4.10.6.

EXERCISE 4.12

1. (ii): Use Lemma 4.12.1.
2. If $(X,Y) = (p,q)$ is a solution in K for $aX^2 + bY^2 + c = 0$, then $K(x,y) = K(t)$ with $t = (x - p)/(y - q)$. The if part can be reduced to this case. The only if part: If $K(x,y) = K(t)$, then x, y are expressed as ratios of polynomials in t; $x = f(t)/g(t)$, $y = h(t)/k(t)$. If $g(s)k(s) \neq 0$ with $s \in K$, then $(f(s)/g(s), h(s)/k(s))$ is a solution in K for $aX^2 + bY^2 + c = 0$. Assume the nonexistence of such s. We may assume that $g(t) = t^u g_1(t)$, $k(t) = t^v k_1(t)$, $u \geq v$, $f(0)h(0)g_1(0)k_1(0) \neq 0$. Then $a(f(t)/g_1(t))^2 + b(h(t)t^{u-v}/k_1(t))^2 + ct^{2u} = 0$. Therefore, if $u \geq 0$ then $t = 0$ gives a nontrivial solution for $aX^2 + bY^2 + cZ^2 = 0$; if $u < 0$, then $(0, 0, 1)$ is a solution.
3. $K(z)$ is a C_1-field because K is algebraically closed. Use Exercises 4.12.2 above.
4. If $n = t$, then the assertion is obvious, and we assume that $n > t$. By virtue of Exercise 4.9.1, there are linear combinations

$z_1, \ldots, z_{n-t}$ of $x_1, \ldots, x_n$ with coefficients in K so that $L[x_1,\ldots,x_n]$ is integral over $L[z_1,\ldots,z_{n-t}]$. Then there are elements $a_1, \ldots, a_s$ of L such that (i) $L = K(a_1,\ldots,a_s)$ and (ii) $x_1, \ldots, x_n$ are integral over $K[a_1,\ldots,a_s,z_1,\ldots, z_{n-t}]$. Take $c \in K$ such that, in the rational expressions of $a_1, \ldots, a_s$ in $x_1, \ldots, x_n$, the denominators are not divisible by $z_1 - c$. Set $V = K[x_1,\ldots,x_n]_P$ with $P = (z_1 - c)K[x_1,\ldots, x_n]$. Then $a_i \in V$, and V contains $K[a_1,\ldots,a_s,x_1,\ldots,x_n]$. Set $P' = PV \cap K[a_1,\ldots,a_s,x_1,\ldots,x_n]$, $P'' = PV \cap K[a_1,\ldots,a_s,z_1, \ldots z_{n-t}]$. Then $\text{trans.deg}_K V/PV = n - 1$, V/PV is the field of fractions of $R = K[a_1,\ldots,a_s,x_1,\ldots,x_n]/P'$ and R is integral over $S = K[a_1,\ldots,a_s,z_1,\ldots,z_{n-t}]/P''$. Set $T = K[a_1,\ldots,a_s]/(P'' \cap K[a_1,\ldots,a_s])$. Since $z_1 - c \in P''$, $\text{trans.deg}_T S \leq n - t - 1$, and $\text{trans.deg}_K T \geq t$. This shows that $P'' \cap K[a_1,\ldots,a_s] = 0$ and L can be imbedded as a subfield of V/PV which is a purely transcendental extension of K.

CHAPTER 5

EXERCISE 5.1

1. Note that if v is a valuation, then it must be nonarchimedean, and therefore it must hold that $vf > vg$ implies $v(f + g) = vf$. v is a valuation if and only if $r \geq 1$.

EXERCISE 5.2

1. Adapt the proof of Theorem 5.2.1.

EXERCISE 5.4

3. Let G be the additive group of real numbers, G' that of rational integers, x an irrational number, and F the subgroup generated by x.

EXERCISE 5.5

3. (I): Note that $a,b \in R$, $a - b \in q^sR$ $(s \geq 1)$ imply $a^p - b^p \in q^{s+1}R$ and $a^{p^n} - b^{p^n} \in q^{s+n}R$. (II): Since this part is too difficult, we omit the proof and we recommend the reader to look at some literature containing a detailed exposition on Witt

vectors (for instance, Jacobson, Lectures on Abstract Algebra, Vol. 3, North Holland). Additional comment: In general, a Witt ring is defined, starting with an arbitrary ring K of characteristic p (prime) and using f_n, g_n in (II). The Witt ring of infinite length is a discrete valuation ring if and only if K is a perfect field.

4. For $m_i = \sum_{j=1}^n a_{ij}u_j$ ($a_{ij} \in K$), it suffices to show that $m_1, \ldots, m_i, \ldots$ converges to 0 if and only if each $\{a_{ij} \mid i = 1, 2, \ldots\}$ converges to 0. The if part follows easily from $|\sum b_ju_j| \leq \sum|b_ju_j| = \sum (vb_j)|u_j|$. The only if part: Assume that $\{a_{i1}\}$ does not converge to 0. Then taking a subsequence, we may assume that $v(a_{i1}) > r$ for a positive number r and all i. Then, considering the sequence $\{a_{i1}{}^{-1}m_i\}$, we may assume that $a_{i1} = 1$ for all i. Then, we apply induction on n to $\{m_{t(i)} - m_i \mid i = 1, 2, \ldots\}$ with an arbitrary increasing sequence $t(1) < t(2) < \cdots$, and we see that each $\{a_{ij} \mid i = 1, 2, \ldots\}$ converges to some element, say a_j^*. Then $u_1 + a_2^*u_2 + \cdots + a_n^*u_n = 0$.

EXERCISE 5.6

1. Let B be a transcendence base of C over Q. Then $\#(B) = \#(B \cup \{t\})$, and there is an injection of $C(t)$ into C.
2. A field K has no archimedean valuation if and only if either $\#(K) > \#(R)$ or the characteristic of K is different from 0.

EXERCISE 5.7

1. Let a be a nonunit in R such that a is not in a proper prime ideal P, and let b be an element of P. Then $a^{-n}b \in P$ for any natural number n.
2. The intersection of some normal rings in normal.
4. The last assertion: Let a, b be elements of R. $wa = wb$ if and only if $aS = bS$, or equivalently, $aR = bR$.
6. For an isolated subgroup H, take $P = \{a \mid va > \text{any element of } H\}$.
7. (1) is easy by virtue of Exercise 5.7.1. As for (2), let $0 = P_0 \subset P_1 \subset \cdots \subset P_n$ be the prime ideals, and let a_i be an ele-

ment of P_i which generate $P_iR_{P_i}$ for $i > 0$. Then, for each b of the field of fractions, $bR = a_1^{m_1} \cdots a_n^{m_n} R$ with rational integers m_i.

8. Prove (1) by induction on n, and then apply it to (2).
9. Let $x_1, \ldots, x_m \in R$ be such that (i) $vx_1 = \cdots = vx_r = 0$ and $x_1 \bmod P_n, \ldots, x_r \bmod P_n$ are algebraically independent over R/P_n and (ii) $vx_{r+1}, \ldots, xv_m$ are linearly independent over Z. If $x_1, \ldots, x_m$ are algebraically dependent over F, then $\sum c_{i_1 \cdots i_m} x_1^{i_1} \cdots x_m^{i_m} = 0$ with coefficients in F. We consider the sum of those terms with least value, say h, under v, we denote the sum by $\sum^* c_{i_1 \cdots i_m} x_1^{i_1} \cdots x_m^{i_m}$. It has value greater than h. In this sum, $(i_{r+1}, \ldots, i_m)$ must be common to all terms, and we see algebraic dependence of $x_1, \ldots, x_r$ modulo P_n.

EXERCISE 5.8

1. The condition is that $i \neq j$ implies that R_i and R_j generate K.

EXERCISE 5.9

1. The case where K is the prime field is the completion of the valuation ring Z_{pZ}. Now, if R is a valuation ring satisfying (i) and (ii), then: (1) If B is an algebraically independent set over R, then consider the polynomial ring P in elements of B over R and construct $R' = P_{pP}$. Then R' satisfies (i) and (ii) and R'/pR' is a purely transcendental extension of R/pR. (2) If a' is an element of the algebraic closure of R/pR, then consider a monic polynomial $f(X)$ over R such that $f(X)$ modulo pR is the minimal polynomial for a'. Then $R'' = R[X]/f(X)R[X]$ is a valuation ring satisfying (i) and (ii) and $R''/pR'' = (R/pR)(a')$. By virtue of (1) and (2), we see the existence of a valuation ring satisfying (i), (ii), and (iv). Then its completion is the required ring.

EXERCISE 5.10

1. First prove the assertion for $L^* = \mathbb{Q}$ or $K(X)$. Then use the fact that for each valuation w of L^*, there are only a finite number of members of V, V' whose restrictions to L^* are equivalent to w. Use also the approximation theorem or the independence theorem of valuations.

EXERCISE 5.11

1. If the integral closure R' of R in the algebraic closure has at least two maximal ideals, say M, M', then since R_P is a Hensel ring, M, M' contains a unique prime ideal P' lying over P and R/P cannot be a Hensel ring.
3. Adapt our proof of Theorem 5.11.6.
4. The first assertion: Use Theorem 5.11.15. The last question: $K((X))$, $K[[X]]$, and composites of $K[[X]]$ with Hensel valuation rings of K.

EXERCISE 5.12

1. Take a valuation ring V_x of $K(x_1,\ldots,x_n)$ having prime ideals $P_1 \supset P_2 \supset \cdots \supset P_n \supset 0$ such that $P_i = x_i(V_x)_{P_i}$, and containing K. Then apply Theorem 5.12.1 to V_x and $V_x \cap k'(x_1, \ldots, x_n)$.

CHAPTER 6

EXERCISE 6.1

3. 1 is positive and every natural number is positive.
4. Cf. the proof of Theorem 5.7.5.
6. Use Exercise 6.1.5.

EXERCISE 6.2

2. Use Lemma 6.2.7.
3. If part: Prove that the field L^* generated by square roots of positive elements of K over L is formally real, by adapting the proof of Lemma 6.2.1.

EXERCISE 6.3

2. Adapt the proof of Theorem 6.3.1.

EXERCISE 6.4

1. In general, if M is the maximal ideal of a valuation ring R of a field K, then $R = \{x \in K \mid xM \subseteq M\}$; this is easily proved by using a valuation defined by R.
2. Adapt the proof of Theorem 6.4.2.

CHAPTER 7

EXERCISE 7.1

3. (2): Prove first that if F' is a finite extension of F, then $G(F'/F)$ is generated by σ such that $x^\sigma = x^q$.

EXERCISE 7.3

2. (1): Let x, y be algebraically independent elements over a field k, and set $K = k(x,y)$, $R = k[x,y]$. Let z be such that $z^2 + xz + y = 0$, and set $K^* = K(z) = k(x,z)$, $R^* = k[x,y,z] = k[x,z]$. Then an example against (i) is: $P^{**} = (x - 1)R^*$, $P^* = (x - 1)R^* + zR^*$. An example against (ii) is: $P^{**} = zR^*$, $P^* = xR^* + zR^*$.

LIST OF SYMBOLS

Symbol	meaning	Section
$\mathbb{C}$	the complex number field	2.1
$\mathbb{Q}$	the rational number field	2.1
$\mathbb{R}$	the real number field	2.1
$\mathbb{Z}$	the ring of rational integers	2.1
Aut	automorphism group	2.2 (cf. Chap. 1)
deg	degree	2.5
Der( )	module of derivations	4.3
det	determinant	
$\frac{d}{d}$, $\frac{\partial}{\partial}$	($\frac{df(X)}{dX}$, $\frac{\partial f(X,Y)}{\partial X}$ etc.) derivations	4.3
G( )	Galois group	3.5, 4.7
Hom( )	set of homomorphisms	2.2 (cf. Chap. 1)
ht	height	4.8
i( )	order of inseperability	4.3
J( )	jacobian matrix	4.3
Krull dim	Krull dimension	4.8
trans.deg	transcendence degree	4.1
$\cong$	isomorphic	2.2 (cf. Chap. 1)
$\oplus$	direct sum	2.2 (cf. Chap. 1)
$\otimes$	tensor product	4.2
$\times$	local tensor product (cf. direct product)	4.2
$\vee$	generation	3.1 (cf. Chap. 1)
( )	(K(X), etc.) field generation	3.1
[ ]	(R[X], etc.) ring generation or polynomial ring	2.5

INDEX